Tobias Senn

Process development for nanostructuring and 3D micro/nanointegration

Tobias Senn

Process development for nanostructuring and 3D micro/nanointegration

Südwestdeutscher Verlag für Hochschulschriften

Impressum / Imprint
Bibliografische Information der Deutschen Nationalbibliothek: Die Deutsche Nationalbibliothek verzeichnet diese Publikation in der Deutschen Nationalbibliografie; detaillierte bibliografische Daten sind im Internet über http://dnb.d-nb.de abrufbar.
Alle in diesem Buch genannten Marken und Produktnamen unterliegen warenzeichen-, marken- oder patentrechtlichem Schutz bzw. sind Warenzeichen oder eingetragene Warenzeichen der jeweiligen Inhaber. Die Wiedergabe von Marken, Produktnamen, Gebrauchsnamen, Handelsnamen, Warenbezeichnungen u.s.w. in diesem Werk berechtigt auch ohne besondere Kennzeichnung nicht zu der Annahme, dass solche Namen im Sinne der Warenzeichen- und Markenschutzgesetzgebung als frei zu betrachten wären und daher von jedermann benutzt werden dürften.

Bibliographic information published by the Deutsche Nationalbibliothek: The Deutsche Nationalbibliothek lists this publication in the Deutsche Nationalbibliografie; detailed bibliographic data are available in the Internet at http://dnb.d-nb.de.
Any brand names and product names mentioned in this book are subject to trademark, brand or patent protection and are trademarks or registered trademarks of their respective holders. The use of brand names, product names, common names, trade names, product descriptions etc. even without a particular marking in this work is in no way to be construed to mean that such names may be regarded as unrestricted in respect of trademark and brand protection legislation and could thus be used by anyone.

Coverbild / Cover image: www.ingimage.com

Verlag / Publisher:
Südwestdeutscher Verlag für Hochschulschriften
ist ein Imprint der / is a trademark of
OmniScriptum GmbH & Co. KG
Heinrich-Böcking-Str. 6-8, 66121 Saarbrücken, Deutschland / Germany
Email: info@svh-verlag.de

Herstellung: siehe letzte Seite /
Printed at: see last page
ISBN: 978-3-8381-5028-4

Zugl. / Approved by: Berlin, TU, Diss., 2012

Copyright © 2015 OmniScriptum GmbH & Co. KG
Alle Rechte vorbehalten. / All rights reserved. Saarbrücken 2015

I

'Everything should be made as simple as possible but not simpler'
(Albert Einstein)

Acknowledgements

First of all, I gratefully thank my supervisor Prof. Dr. Martin Schmidt for all his help and the inspiring discussions during these last 3 years.

I would like to express my gratitude to Dr. Helmut Schift for his help and advice, especially during the writing phase. His treasure trove of experience and the fruitful discussions helped a lot in the interpretation of the experimental results.

Prof. Dr. Alexei Erko, head of the Institute for Nanometre Optics and Technology, I thank for the continuing support during the accomplishment of this thesis.

A special thanks goes to Dr. Bernd Löchel for his great support and advice. His cheerfulness and optimism always motivated me to do my best.

I thank Dr. Marcus Lörgen for his support, especially at the beginning of my time at the HZB. His motivating words and initiative helped greatly in finding the final topic of this thesis.

I would like to thank Dr. Arne Schleunitz for his help during the training at the beginning and the ongoing discussions about the world of NIL.

I thank my colleges Christian, Christoph, Johannes, Ivo, Daniel, Maha, Olli, Antje, Harald, Alexander, Tino, Thomas, Tom, Adrian and Janice from the AZM group for the nice working atmosphere and the support.

Also I want to thank the students Janika, for conducting the NIL experiments and Philipp for his help with the paper and thesis correction. Their enthusiasm and their attention to detail helped to improve the experimental results and the standard of English in the publications and this thesis. I want to thank Christoph, Alexander, Thomas and Heike for their help in many different areas.

I would also like to thank my PhD colleges Nils, Steffie, Max, Jürgen and Jens for the good collaboration in different projects and the nice and inspiring conversations.

A very special thanks goes to Dr. Neus Sabaté and Dr. Juan Pablo Esquivel Bojorquez from the CNM in Barcelona for the great and fruitful collaboration. The impact of this collaboration on the final form of my thesis can be seen clearly.

I thank my family, my parents and my sister, for all their support over the last years and their helpful advice in the difficult decisions in my life. It is to them that I dedicate this work.

Finally, I thank Ana for being so patient and understanding during my long labor and for always being there for me.

List of Publications

Journal Paper

Integration of moth-eye structures into a PDMS stamp for the replication of functionalized microlenses using UV-NIL
T. Senn, O. Kutz, C. Weniger, J. Li, M. Schoengen, H. Löchel, J. Wolf, P. Göttert and B. Löchel, J. Vac. Sci. Technol. B 29(6), Nov/Dec 2011 (**This article has been selected for the October 17, 2011 issue of Virtual Journal of Nanoscale Science & Technology.**)

Fuel cell-powered microfluidic platform for Lab-on-a-Chip applications
J.P. Esquivel, M. Castellarnau, T. Senn, B. Löchel , J. Samitier and N. Sabaté, Lab Chip, DOI: 10.1039/c1lc20426b. (2011)

Fabrication of photonic crystals for applications in the visible range by Nanoimprint Lithography
T. Senn, J. Bischoff, N. Nüsse, M. Schoengen, B. Löchel, Photonics and Nanostructures – Fundamentals and Applications 9 (2011) 248–254

Fabrication of high aspect ratio nanostructures on 3 D surfaces
T. Senn, J.P. Esquivel, N. Sabaté, B. Löchel, Microelectron. Eng. 88 (2011) 3043–3048

3D structuring of polymer parts using thermoforming processes
T. Senn, Ch. Waberski, J. Wolf, J.P. Esquivel, N. Sabaté, B. Löchel, Microelectron. Eng. 88 (2011) 11–16

Towards a compact SU-8 micro direct methanol fuel cell
J.P. Esquivel, T. Senn, P. Hernández-Fernández, J. Santander, M. Lörgen, S. Rojas, B. Löchel, C. Cané, N. Sabaté, J. Power Sources 195 (2010) 8110–8115

Replica molding for multilevel micro–\nanostructure replication
T. Senn, J. P. Esquivel, M. Lörgen, N. Sabaté and B. Löchel, J. Micromech. Microeng. 20 (2010) 115012 (8pp)

Replication of HARMST and large area nanostructured parts using UV cationic polymerization
T. Senn, C. Mueller and H. Reinecke, J. Micromech. Microeng. 20 (2010) 075002 (6pp)

Conference Paper

Frequency tuning of ultrafast magnetization oscillations by varying the iron content of FePt alloys
R. Brandt, F. Ganss, T. Senn, M. Albrecht, H. Schmidt, 56th Annual MMM Conference, 30.10.11-03.11.11, Scottsdale, Arizona, USA

Shape dependent magnetization dynamics in single FePt nanomagnets
R. Brandt, C. Brombacher, D. Gilbert, P. Krone, F. Ganss, T. Senn, K. Liu, M. Albrecht, H. Schmidt, 56th Annual MMM Conference, 30.10.11-03.11.11, Scottsdale, Arizona, USA

Soft lithography method for the integration of surface structures into polymer devices
T. Senn, C. Weniger, J. P. Esquivel, D. Schondelmaier, N. Sabaté, B. Löchel, MNE, 19.09.11-23.09.11, Berlin, Germany **Invited Speaker**

Nanostructuring of 3D shaped thermoplastic foils for the fabrication of functionalized optical and fluidic devices
T. Senn, C. Weniger, J.P. Esquivel, N. Sabaté, B. Löchel, 5th MPA Meeting, 27.06.11-29.06.11, Alvor, Portugal

Functionalization of 3D shaped microchannels with high aspect ratio nanostructures
T. Senn, C. Weniger, J. P. Esquivel, N. Sabaté, B. Löchel, 9th international Workshop on High Aspect Ratio Microstructures (HARMST), 12.06.11-18.06.11, Taiwan, **Best Student Oral Presentation Award**

Prozessoptimierung zur Herstellung von zweidimensionalen photonischen Kristallen mittels Nanoimprint Lithografie
J. Bischoff, T. Senn und B. Löchel, GMM Workshop Mikro- Nanointegration, 03.03.11-04.03.11, Stuttgart

A compact and fully-integrated SU–8 micro direct methanol fuel cell
J.P. Esquivel, T. Senn, J. Santander, P. Hernández–Fernández, S. Rojas, M. Lörgen, N. Torres-Herrero, I. Gràcia, E. Figueras, C. Cané, B. Löchel and N. Sabaté, Power-MEMS, 30.11.10-03.12.10, Leuven, Belgium

Nanoprägen in Polymerfolien als Fertigungsverfahren für fluidische Mikro-Nano-Systeme
A. Schleunitz, T. Senn, P. Datta, J. Göttert, S. Giselbrecht, M. Reinhardt, B. Löchel, MikroSystemTechnik - KONGRESS 2009, 12.10.09-14.10.09, Berlin

Patent Applications

Method of producing a polymer stamp for the reproduction of devices comprising microstructures and nanostructures, a corresponding polymer stamp, and a corresponding device
T. Senn, J.P. Esquivel, N. Sabaté, M. Lörgen, EP 10194892.5 2010

Method of producing a polymer stamp for the reproduction of devices comprising microstructures and nanostructures, a corresponding polymer stamp, a corresponding device and a fuel delivery system or water management system for fuel cells
J.P. Esquivel, T. Senn, M. Lörgen, N. Sabaté, C. Cané, EP 10153537.5 2009

Abstract

Micro- and nanostructures significantly influence the physical and chemical properties of surfaces. Therefore, such structures can be used to functionalize surfaces and adjust the surface properties to the specific requirements of a certain application. A typical surface effect based on topological patterning is the self-cleaning surface, similar to the well-known lotus effect based on micro- and nano roughness. Other examples are the effect of the moth eye to create an anti-reflective effect, based on sub-wavelength gratings; the enhancement of surface adhesion similar to the Gecko's ability to climb on smooth surfaces due to large hairy nanostructures; or the effect of the shark skin where a microstructure influences the flow resistance and spore settlement. These properties as well as many others can have an impact on industrial applications in different fields. Many approaches have been proposed for the fabrication of such structures to imitate the effects of nature and use them in industrial applications. For economical use of these structure effects, fabrication processes are needed that meet the standard set by industry: the production of robust structures with a high lifetime in an easily reproducible and cost efficient way. Typical replication processes to fulfil these requirements are Injection Molding or Imprint processes. A remaining challenge is the integration of functional structures into microdevices to achieve e.g. patterning in 3D, 3D surface structures and sidewall patterning.

The goal of this work is to fabricate and replicate functional surface structures using Imprint technologies and to integrate them into polymer parts to obtain a higher functionality of polymer based microdevices. For this purpose, a Nanoimprint Lithography (NIL) process was developed to replicate nanostructures in the sub 100 nm range. The potential of this technology is demonstrated in various applications including the replication of photonic crystal structures for the fabrication of novel optical devices and the structuring of functional materials for applications in the fields of solar cells and magnetic sensors.

For the integration of functional surface structures into polymer based microparts, a process chain was developed based on Thermoforming and Soft Lithography. In this process chain a 3D structured elastomer stamp is produced from which replications can be made.

This process chain was applied to various industrial fields like optics and micro fluidics. Functionalized micro lenses were produced by the integration of moth eye structures; in the field of micro fluidics, a hydrophobic surface structure was molded into micro channels to create a fluidic system for a micro fuel cell. The results of these investigations have shown that the integration of surface structures into polymer parts can be used to improve the performance of a system and to reduce production costs.

Kurzfassung

Mikro- und Nanostrukturen haben Einfluss auf die physikalisch-chemischen Eigenschaften von Oberflächen. Dadurch ist es möglich die Oberflächeneigenschaften durch eine Strukturierung an spezifische Anforderungen einer Anwendungen anzupassen. Ein typischer durch Strukturen hervorgerufener Oberflächeneffekt ist eine selbstreinigende Oberfläche, ähnlich dem bekannten Lotuseffekt basierend auf einer Mikro- und Nanorauhigkeit. Andere Beispiele sind der Effekt der Mottenaugen um Antireflektionseigenschaften basierend auf einem Subwellenlängengitter zu erzeugen, die Erhöhung der Oberflächenadhäsion durch langhaarige Nanostrukturen (Gekko-Effekt), oder der Effekt der Haifischhaut, bei der eine Mikrostruktur den Strömungswiderstand und die Anlagerung von Bakterien beeinflusst. Für die Herstellung und das Design künstlicher Strukturen, die diese aus der Natur bekannten Eigenschaften imitieren, wurden unterschiedliche Ansätze entwickelt. Um das wirtschaftliche Potential dieser Effekte zu nutzen, benötigt man Herstellungsprozesse, welche die Anforderungen der Industrie erfüllen: Die reproduzierbare und kosteneffiziente Produktion robuster Strukturen mit hoher Lebenszeit. Typische Replikationsprozesse, die diese Anforderungen erfüllen sind Spritzgieß- und Prägeprozesse. Eine weiterhin bestehende Herausforderung ist die Integration dieser Strukturen in Mikrobauteile.

Ziel dieser Arbeit ist es funktionale Oberflächenstrukturen mittels Imprint Prozessen zu replizieren und sie in Polymerbauteile zu integrieren, um hoch funktionalisierte Mikroteile herzustellen. Hierzu wurde zunächst ein Nanoimprint Lithographieprozess entwickelt, um Nanostrukturen im sub 100 nm Bereich zu replizieren. Das Potential dieses Verfahrens wird in mehreren Anwendungen demonstriert, unter anderem die Replikation photonischer Kristalle für die Herstellung neuartiger optischer Anwendungen und die Strukturierung funktionaler Materialien für Anwendungen im Bereich der Solarzellen- und Magnetsensorforschung.

Für die Integration funktionaler Oberflächenstrukturen in Polymerbauteile wurde eine Prozesskette entwickelt, basierend auf einem Thermoform- und Softlithographieprozess. Hierbei wird ein dreidimensional strukturierter Elastomerstempel hergestellt, welcher in einem Gießprozess verwendet wird um Polymerbauteile zu produzieren.

Mit Hilfe der entwickelten Prozesskette wurden mehrere Bauteile für optische und mikrofluidische Anwendungen gefertigt. Durch die Integration einer Mottenaugenstruktur in ein Linsenarray konnten funktionalisierte Linsen hergestellt werden. In einer mikrofluidischen Anwendung wurden hydrophobe Oberflächenstrukturen in Mikrokanäle integriert, um so ein Befüllungs- und Abgassystem für eine Mikrobrennstoffzelle herzustellen. Die Ergebnisse dieser Untersuchungen haben gezeigt, dass die Integration von funktionalen Oberflächenstrukturen in Polymerbauteile dazu genutzt werden kann, die Eigenschaften eines Systems zu verbessern und die Kosten im Herstellungsprozess zu reduzieren.

Contents

Abstract	IX
Kurzfassung	XI
Contents	XIII
1 Introduction	**1**
2 Fundamentals and Methods	**5**
2.1 Nanoimprint Lithography	6
2.1.1 Process	6
2.1.1.1 Hot Embossing Nanoimprint Lithography (HE-NIL)	6
2.1.1.2 Ultraviolet Nanoimprint Lithography (UV-NIL)	9
2.1.2 Materials	12
2.1.2.1 PMMA as Molding Material	12
2.1.2.2 UV-Curable Materials	13
2.1.3 Comparison	14
2.1.4 Flow Behaviour of Thin Film	16
2.2 Thermoforming	19
2.2.1 Classic Thermoforming Process	19
2.2.1.1 Methods of Thermoforming	19
2.2.1.2 Foil Heating Mechanism	20
2.2.1.3 Foil Stretching	20
2.2.1.4 Material Aspects	21
2.2.2 Micro–Thermoforming	22
2.3 Soft Lithography	23
2.3.1 PDMS as stamp material for Soft Lithography	24
3 Stamp Fabrication	**29**
3.1 Stamps for HE-NIL	29
3.1.1 Fabrication Process for Si Stamps	29
3.1.2 Anti-sticking Layer for the Stamps	31
3.2 Stamps for UV-NIL	32
4 Replication of 2D Nanostructures	**35**
4.1 Imprint Machine HEX03	35
4.2 HE-NIL Process	35
4.2.1 Influence of Structure Size and Form on Residual Layer	37

5	**Applications of HE-NIL**		**39**
	5.1 Replications of Photonic Crystals		39
		5.1.1 Photonic Crystals	39
		5.1.2 Optical Properties of Different 2D Photonic Crystals	40
		5.1.3 Photonic Crystal Design	40
		5.1.4 Replication of Photonic Crystal with NIL	43
		5.1.5 Replication Results	45
	5.2 Outlook		48
		5.2.1 Structuring of Substrates for Applications in Solar Cells	49
		5.2.2 Structuring of Metallic Layers	50
6	**Fabrication of 3D Structures**		**55**
	6.1 Motivation and Process		55
	6.2 Thermoforming: Equipment and Process		55
		6.2.1 Equipment	55
		6.2.2 Thermoforming Results	57
	6.3 Foil Structuring using Hot Embossing		60
		6.3.1 Process	60
		6.3.2 Thermoforming Results using Hot Embossing for Foil Structuring	61
		6.3.3 Extending the Boundary Conditions	62
	6.4 Foil Structuring using UV-NIL		64
		6.4.1 Process	64
		6.4.2 Thermoforming Results using UV-NIL for Foil Structuring	66
	6.5 Possibilities and Limits		70
		6.5.1 Foil Thickness Distribution	70
		6.5.2 Geometrical Limitations	72
7	**Replication of 3D Structures using Replica Molding**		**79**
	7.1 Replica Molding for Multilevel Micro- Nanostructure Replication		79
	7.2 3D Structuring of Polymer Parts		84
8	**Applications for 3D Structured Polymerparts**		**89**
	8.1 Anti-Reflection Structures for Microlenses		89
		8.1.1 Foil Structuring	90
		8.1.2 Fabrication of the lens array and replication of functionalized lenses	91
		8.1.3 Replication Results and Conclusion	92
	8.2 Fuel Delivery System for Micro Fuel Cell Applications with Passive Feed		96
		8.2.1 Hybrid Fuel Cell and System Integration	97
		8.2.2 First Version of the Fuel Delivery System	98
		8.2.2.1 Fabrication Process	100
		8.2.2.2 Fluidic Characterization and Discussion	101

8.2.3		Second Version of the Fuel Delivery System	103
8.2.4		Final Version of the Fuel Delivery System	105
	8.2.4.1	Chemical Treatment of the Structured Surfaces and Characterization	105
	8.2.4.2	Fluidic Characterization	107

9 Summary and Outlook   **111**
   9.1 Summary 111
   9.2 Outlook 114

Abbreviations   **115**

Formula Symbols   **117**

List of Tables   **119**

List of Figures   **121**

References   **127**

Appendix   **141**

1 Introduction

Since the visionary talk entitled 'There's Plenty of Room at the Bottom' given by Richard P. Feynman at the Annual meeting of the American Physical Society in 1959, where he presented the vision of extreme miniaturization, the technological progress has led to many inventions and developments in the fabrication of nano-objects. Especially in the last 15 years, nanotechnology has gained importance in science and engineering and it is likely to have a profound impact on our economy and society in the early 21st century. Nanotechnology is the ability to manipulate individual atoms and molecules to produce nanostructured materials and submicron objects [1].

Micro- and nanostructures have influence on the physical and chemical properties of surfaces and can therefore be used to control them. These include optical, adhesive or wetting properties. Many functional surface structures can be adapted from nature which is the focus of the field of technology called bionics. One example in the field of optics are photonic crystals [2], periodically structured dielectric materials, which are responsible for the colors of butterfly wings, opal stones and the peacock's feathers. These natural structures have inspired scientists to further investigations leading to the development of novel photonic elements like filters, splitters or sharped bend waveguides. Another example in the field of optics are the structures from the moth's eye. Here a regular structure in the sub-wavelength range results in a gradual change of the refractive index from one medium to another causing an anti-reflective effect [3]. Such structures have been successfully applied to applications in the field of solar cells and displays [4]. But not only optical properties can be influenced by surface structures, the most famous surface effect based on micro- and nanostructures is the lotus effect [5]. Here the combination of micro and nanostructures together with a wax coating on the surface results in superhydrophobic wetting properties. Using such structures, self-cleaning and drag reduction surface properties can be obtained.

In order to use the commercial potential of such bio-inspired surface structures, technological solutions are needed for the economical fabrication and replication [6]. In recent years, great efforts have been put into the fabrication and replication of nanostructures. Many approaches have been developed and improved in order to make these technologies available for industrial applications. One of the most promising technologies in the field of nanostructure replication is Nanoimprint Lithography (NIL) [7]. The constant improvement and further technological development in the last years have made this technology a powerful tool for nanoreplication. The main difference between patterning technologies based on imprint and those based on photolithographic methods is that imprint technologies mechanically deform a resist material, whereas in photolithographic methodology the resist is structured by selective exposure. Therefore, the resolution that can be obtained using these technologies is not limited by light diffraction or beam scattering [8]. Furthermore, these imprint technologies require little capital investment compared to other technologies capable of achieving comparable resolutions, like electron beam writing (EBL) or X-Ray Lithography, making them attractive replication techniques for industrial applications.

Besides the fabrication and replication of these functional structures, the integration into micro or macro systems creates further challenges. By integrating a structure based anti-reflective layer into a micro lens array, the additional process steps that are typically necessary to obtain a blooming of micro lenses can be avoided which reduces production costs. Additionally these anti-reflective layers can function in harsh environment, since critical material interfaces can be avoided. The integration of functional surface structures can also be used to improve existing systems or create completely new products. The specific integration of, for example, hydrophobic areas into micro fluidic devices can be used for gas-liquid separation or the reduction of flow resistance. It has been demonstrated that Soft Lithography processes are capable of replicating complex structures with high fidelity [9]. In order to replicate such structures, master structures are needed from which replications can be made.

The goal of this work is two-sided: on one side, the replication of functional structures on 2D surfaces and, on the other side, the integration of these functional structures into micro or macrostructures to fabricate functional devices. The replication of 2D structures and the structuring of functional metal layers is done using NIL processes. In order to replicate structures using NIL, master structures must initially be fabricated. The stamps for the NIL process are fabricated using Electron Beam Lithography (EBL), a typical fabrication process for nanostructures.

For the integration of functional structures into micro or macrostructures a process chain was developed, starting with the structuring of a thermoplastic foil, which is done either by Hot Embossing or Ultraviolet NIL (UV-NIL). These structured foils are molded over a structured insert via a Thermoforming process to cover a structured mold insert. In this way, the structures on the thermoplastic foil are integrated into the microstructures of the mold insert. This structured foil is further used as a master structure for a Soft Lithography method. In this method, an elastomer material is used to produce a soft stamp from the structured foil. These stamps are used in an UV based casting process (Replica Molding process) for the fabrication of functionalized polymer devices as well as for UV-NIL for the replication of 3D structured polymer films.

1. Introduction

This work is divided into 7 main chapters:

Chapter 2 describes the technological background and the current state of the fabrication methods applied in this work. This includes the basics and the variations of NIL, material aspects and a comparison of the different processes. Furthermore, the Thermoforming process, which is used for the structure integration and master fabrication, is explained and in the last section of this chapter, an overview of Soft Lithography methods and a detailed explanation of the Replica Molding process are given. Additionally, typical characteristics of Soft Lithography methods based on the materials aspects are discussed.

Chapter 3 gives an overview on the techniques used for the fabrication of the stamps. Since two types of stamps were used in this work (hard and soft stamps), the first section describes the fabrication process for hard stamps used for Hot Embossing NIL (HE-NIL) while the second section details the fabrication of the soft stamps which were used for UV-NIL.

Chapter 4 starts with a presentation of the Hot Embossing machine used in this work. Then the development of the process steps for HE-NIL and the replication results are shown. In the last section, the influence of structure size and form on the residual layer distribution are discussed based on experimental results.

Chapter 5 presents the structures actualized using the HE-NIL process. In the first section the replication and characterization of photonic crystal structures is presented. In the second section, the results of the experiments on the structuring of functional materials for applications in novel solar cells and the structuring of metallic layers for catalytic and magnetic applications are shown.

Chapter 6 describes the development of the Thermoforming process, which was used for the integration of functional surface structures into microstructures. Here the design of the Thermoforming tool, the pressure system and the different processes for the foil structuring are presented. Furthermore, the results of the structure integration and the limits of this technology are discussed.

Chapter 7 describes the Replica Molding process which was used in this work for the replication of micro and nanostructures in polymer material. In the first section, multilevel micro and nanostructures were replicated using this process showing the maximum resolution that can be obtained and, in the second section, the replication of 3D structures is described.

Chapter 8 shows the applications actualized using the combination of the Thermoforming and the Replica Molding process. The processes presented in the previous chapters were used to fabricate functionalized micro lenses which could be applied to mobile camera modules, and a functionalized micro fluidic device was produced which was integrated into a micro fuel cell as fuel delivery and out-gassing system.

2 Fundamentals and Methods

This work deals with the replication of 2D nanostructures as well as the fabrication and replication of 3D structured polymer parts. In this chapter, an overview of the current status of the utilized technologies will be given. The fabrication of 2D nanostructures in this work is done by Nanoimprint Lithography (NIL), a Next Generation Lithography (NGL) method, which is believed to be one of the most promising techniques for volume manufacturing of nanostructured components [10]. The most important categories of NIL are Hot Embossing Nanoimprint Lithography (HE-NIL) and Ultraviolet Nanoimprint Lithography (UV-NIL). For simplification and comparison with Hot Embossing, where foils or plates rather than thin resists coated on solid substrates are structured, HE-NIL was chosen as term for the structuring of thin resists on solid substrates. In the first section, a brief explanation of the two NIL processes will be given, followed by a discussion of the material aspects of stamps and resist materials. A decisive factor for the understanding of NIL processes is thin-film rheology as the pattern transfer occurs by mechanical deformation of the resist material. These aspects will be discussed in detail in the following section. The structuring of vertical side walls, which is also a goal of this work, is not possible with rigid stamps typically used in the HE-NIL process. In order to fabricate such 3D structured components and parts, different strategies have to be developed. The term 3D structuring can be used for various structure shapes, for example, patterns containing a 3D relief on a 2D substrate [11, 12]. Such structures can be replicated by the means of NIL processes using rigid stamps [13, 14]. When using soft stamps, 3D patterns are also replicable even if they contain undercuts[15, 16]. 3D structures can also be obtained by forming a 2D substrate over a master to obtain a 3D structured substrate [17]. In this work, a combination of both types of structures was obtained, the structuring of binary resist type structures or a 3D relief on a 2D substrate, and further the integration of these structures into a 3D structured master, using a micro–Thermoforming process. This process also allows the addition of structures to vertical side walls. Micro–Thermoforming is a modification of the classic Thermoforming process, a common process, for example, in the automobile or packaging industry. The basics of the micro–Thermoforming process and its cutting-edge developments will be discussed in section 2.2. The replication of the 3D structured master was done using Replica Molding, a Soft Lithography method. In the last section of this chapter, a brief overview of Soft Lithography processes will be given followed by a detailed explanation of the Replica Molding process. In this respect, potentials and limits of the technology, which are especially attributed to material properties, will be discussed.

2.1 Nanoimprint Lithography

Nanoimprint Lithography is a replication technique which offers high resolutions compared to other technologies. It was demonstrated that structure details in the sub 10 nm range can be replicated using this technology [18, 19]. Since the first publications [7, 20, 21], Nanoimprint Lithography has attracted the increasing interest of scientists all over the world. The parallel and fast fabrication of micro and nanostructures using NIL and additionally the high resolution that can be achieved has made it a promising candidate as NGL method [22, 23]. Furthermore, other NGL technologies like EBL or Extreme UV Lithography (EUV Lithography) require expensive machines and therefore high investments. This makes NIL an attractive alternative as low cost replication technique. Once a stamp is fabricated using e.g. EBL, replications from this master can be made in a parallel process. In NIL, the concept of resist patterning is used for structure replication. In contrast to other techniques, optical lithography or EBL, for example, the pattern is replicated due to mechanical deformation of the resist using the difference in the mechanical properties of the structured stamp and the molding material. Therefore, it is possible to achieve resolutions beyond the limitations set by light diffraction or beam scattering that are encountered in conventional techniques like UV-Lithography [8]. The success of NIL, especially as lithography method for microelectronics and hard discs, depends on the ability to solve processing issues such as resolution, defect control, fidelity, overlay, repeatability and throughput [10].

2.1.1 Process

2.1.1.1 Hot Embossing Nanoimprint Lithography (HE-NIL)

In order to achieve a good replication quality of nanostructures with HE-NIL, the development of various process steps is necessary. An overview over the process steps is schematically illustrated in Fig.2.1. A description of the process steps with parameters can further be found in the Appendix (process 3)

In the first step, a stamp is produced. The most common way of stamp fabrication is EBL with a subsequent Reactive Ion Etching (RIE) step to transfer the structures into the substrate (see chapter 3.1.1). This stamp is used to transfer the structures into a thermoplastic material (typically PMMA) by applying temperature and pressure. The HE-NIL process consists of the following process steps. First the stamp and the coated substrate are brought into contact under vacuum condition in order to avoid trapping air in the imprinted layer. Then, the sandwich-like assembly is heated up above the glass transition temperature (T_g) of the polymer layer and pressure is applied. Due to the temperature, the viscosity of the polymer layer is decreased and the cavities in

2. Fundamentals and Methods

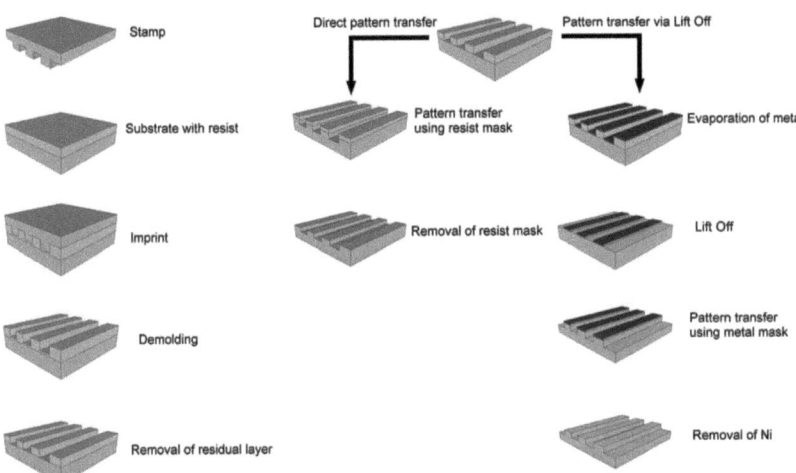

Figure 2.1: Schematic overview over the process steps during NIL: First, the structures in the stamp are transferred into a PMMA layer on a substrate using temperature and pressure. Then the remaining residual layer is removed by RIE (left side). In the following steps the structured PMMA layer is used to transfer the structures into the substrate either by direct pattern transfer (middle) or by transferring the structures via Lift Off (right side)

the stamp can be filled. In the next process step, the stamp and the substrate are separated with a thin residual layer remaining on the substrate (Fig.2.1 left). A typical process sequence during the imprint process is illustrated in Fig.2.2. The sequence starts with the heating of the assembly until the imprint temperature ($T_{imprint}$) is reached (interval 1-2 in Fig.2.2). At this point, pressure is applied and held until the cavities in the stamp are filled (interval 2-3 in Fig.2.2). Once the cavities are filled, the cooling process begins and the pressure is released once ambient temperature is reached (interval 3-4 in Fig.2.2). Because of the variothermal process characteristics of this method and the fact that heating and cooling times are typically high, cycle times depend highly on these two process steps. The temperature and pressure that is needed for a complete cavity filling depends on the resist material. PMMA is the most common resist material but other thermoplastic materials with low T_g like e.g. COC can also be used. Typical imprint parameters and characteristics for HE-NIL can be found in Tab.2.1 (for PMMA typical temperatures range from 170 to 200 °C and the pressure from 5 to 10 MPa).

In order to transfer the structures into the substrate, the residual layer must be removed (Fig.2.1 left). The removal of the residual layer is typically performed using a RIE process with an oxygen plasma. During this process, the resist is homogeneously thinned allowing process windows to be opened; the resist can now be used for further

7

2.1 Nanoimprint Lithography

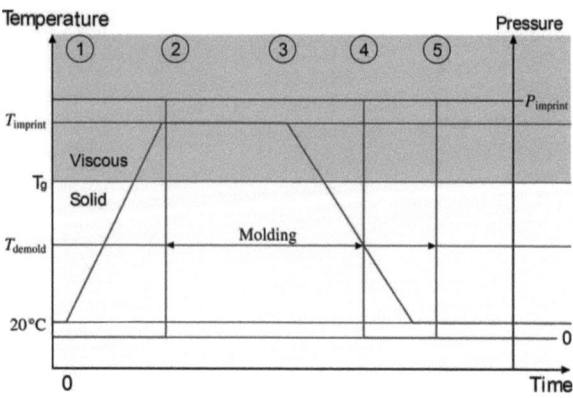

Figure 2.2: Temperature and pressure sequence during the HE-NIL process [10].

processing. Since the oxygen plasma is an non-selective etch process for resist materials, the resist will also be etched away from the sidewalls resulting in a change of the geometry and rounding off the edges. The most typical methods for further processing are using the structured resist as an etch mask (Fig.2.1 middle) or as a mask for a Lift Off process for structuring of functional materials like metals (Fig.2.1 right). Both methods can be used for transferring the structures into the substrate as illustrated in Fig.2.1. For the direct pattern transfer, the resist is used directly as an etch mask in a RIE process. In the pattern transfer via Lift Off, a hard mask is used in the etching process offering the possibility of fabricating higher aspect ratios, as the metal mask has a higher selectivity in the RIE process than the resist. In the Lift Off process, a metal is deposited on the structured resist and the resist is dissolved, removing the metal which is deposited on the resist. This leads to a structured metal mask which can be used in an etching process. Besides using the structured resist as a mask for further processing, the resist pattern can also be used directly as functional layer. In this work, the HE-NIL process was used for structuring the resist as an intermediate mask layer for further structure transfer into the substrate as well as for structuring different metals via Lift Off.

A detailed description of possible applications of functional resist patterns can be found in [24, 25]. The fabrication of functional resist patterns can be improved using roll to roll processes [26]. In roll to roll processes, a thin, "flexible" stamp is wrapped onto a roller surface and fixed by gluing or clamping. In the HE-NIL version of such a roll to roll process, the pattern on the stamp is then transferred onto a thermoplastic foil by applying pressure and temperature while moving the thermoplastic foil under the roller. In this way, pattern transfer can occur quickly with a velocity in the range of m/min [26].

2. Fundamentals and Methods

2.1.1.2 Ultraviolet Nanoimprint Lithography (UV-NIL)

In the UV-NIL process, the pattern transfer occurs, similarly to the HE-NIL process, by mechanical deformation of the resist material. The difference between the two techniques is that in the UV-NIL process a liquid resist, which is cured using UV radiation, is used. In the UV-NIL process, either the substrate or the stamp must be transparent to allow for exposure via UV radiation. Due to the low viscosity of the resist materials typically used for UV-NIL, the mechanical deformation of the material can occur with less pressure compared to HE-NIL. Therefore, soft stamps can be used in UV-NIL. In the following paragraphs, the different variations of the UV-NIL process are discussed. An overview over the process steps and the corresponding parameters as it was performed in this work can also be found in the Appendix (process 4).

UV-NIL using Hard Stamps

The UV-NIL process using hard stamps was first published by Philips Research in 1996 [27]. It was further developed in a so-called Step and Flash Imprint Lithography (SFIL) process by the group of Prof. C.G. Willson [28]. In this publication, an anti-sticking layer was first deposited on a quartz template with a size of 1.5 in. x 1.0 in. and patterns ranging in lateral size from 20 μm to 60 nm were transferred into a resist material which was used as a etch barrier to transfer the pattern into a underneath lying layer. This technique has some advantages over imprinting techniques using thermoplastic resist materials and pattern transfer by applying temperature and pressure. In the SFIL process, the transfer is carried out at room temperature and at low pressures resulting typically in shorter process times (< 1 min). Since the first publication, several process variations have been proposed which mainly differ in the use of stamp materials. The use of rigid quartz templates [29, 30] is normally limited to small areas

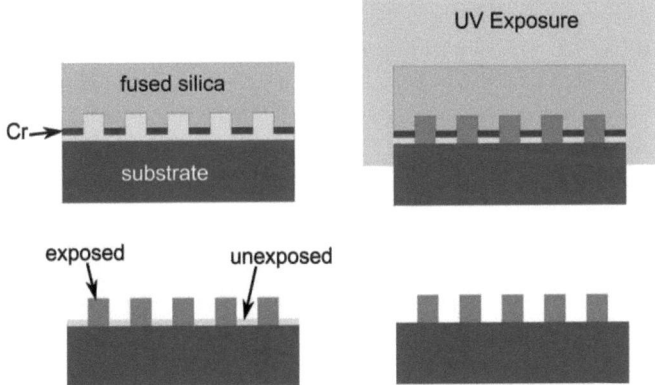

Figure 2.3: Schematic overview for the resist patterning using the CNP process.

9

2.1 Nanoimprint Lithography

because non-uniformities in the substrate cannot be compensated for by the small pressures normally used in this process. Therefore, rigid stamps are typically used in the SFIL process, where small templates with structures down to 20 nm are replicated on a large substrates via repeating steps [31]. The quartz templates are fabricated in the same way as the templates used in the HE-NIL process, only that the Si substrate is exchanged for a quartz substrate. Therefore, the resolutions and limits in structure size are comparable to HE-NIL stamps.

A possibility for allowing replication of large area rigid stamps was proposed by Jeong et al. [32]. Here an elementwise pattered stamp (EPS) was used to replicate nanostructured areas on a 4 in. substrate. The structured elements in the stamp have a size of 13 x 13 mm^2 and are separated by channels. The resist is brought onto the stamp before imprinting, avoiding trapped air in the structured area. Due to the deposition of an anti-sticking layer on the stamp, the structures adhere on the substrate and can be easily released.

One big issue in NIL processes is the controlling of the residual layer. In the UV-NIL process, the residual layer can also be avoided completely using a Combined Nanoimprint and Photolithography (CNP) technique [33]. In this process, a structured Ni or Cr mask is first used as an etch mask for pattern transfer into the quartz stamp. Instead of removing the metal mask, as in other processes, the mask remains on top of the protrusions during the imprint to avoid curing of the resist beneath the mask. In this way, the resist beneath the mask remains unexposed and can be removed in a subsequent developing step. The process steps during this CNP process are summarized in Fig.2.3.

UV-NIL using Soft Stamps

As mentioned before, in imprint processes, the pattern transfer occurs due to the different mechanical properties of the stamp and resist material. Since the resist materials applied to UV-NIL, typically have much lower viscosities compared to those applied to HE-NIL (at $T_{imprint}$), stamp materials suitable for UV-NIL do not have the same mechanical requirements as stamp materials for HE-NIL. Therefore, it is possible to use so called soft stamps [34] in UV-NIL as no or only small pressure has to be applied to obtain cavity filling of the structures. The flexible soft stamps offer various advantages when compared to rigid stamps. Non-uniformities of the substrate can be compensated for by flexible stamp materials allowing larger areas to be structured in a single step.

The fabrication of such flexible stamps is typically done using a casting process (see section 2.2.1) or an imprinting process [15]. In this way various soft stamps can be fabricated from a single master, preventing damages to the expensive master structure. This makes the use of a flexible stamp very attractive from an economical point of view.

2. Fundamentals and Methods

An important factor for a successful pattern transfer using imprint processes, is the uniformity of the residual layer. It has been shown that the use of flexible poly(dimethysiloxane) (PDMS) stamps lead to an extremely good uniformity in thickness of the residual layer with standard deviations in the range of 2-3 nm [35].

UV-NIL processes offer various advantages over HE-NIL processes which explains its increasing use over the past few years. Especially the faster form filling and curing process makes it attractive for roll to roll [36] or roll to plate [37] applications. Roll to roll imprint processes can be used e.g. for the structuring of resist materials on plastic foils [36] in order to functionalize the surface of the foil, whereas roll to plate processes can also be applied for lithography applications to transfer a pattern into the substrate, which has been used for the structuring of solar cells [38]. A detailed discussion of the material aspects of such flexible stamps will follow in chapter 2.2.1. The UV-NIL process has been successfully applied to the fabrication of transistors [39] or optical ring resonators [35] as well as to the fabrication of multilayer structures [40, 41]. In these applications, resolutions down to 80 nm were obtained.

2.1.2 Materials

2.1.2.1 PMMA as Molding Material

One of the most common materials for HE-NIL as well as for EBL is poly(methyl methacrylate) (PMMA). Thermoplastic materials like PMMA offer the possibility of controlling the viscosity by temperature, allowing significant changes of the viscosity in the range of some tens of degrees. One of the most significant influence on physical properties of a thermoplastic material is the molecular weight M_W. Since the development contrast between exposed and unexposed areas increases with M_W, a high-M_W PMMA is used for EBL (typically 600 kg/mol to 2000 kg/mol). The use of PMMA as molding material in NIL requires a strong dependence of the viscosity on the temperature, which can be obtained by a low-M_W PMMA (typically 25 to 100 kg/mol).

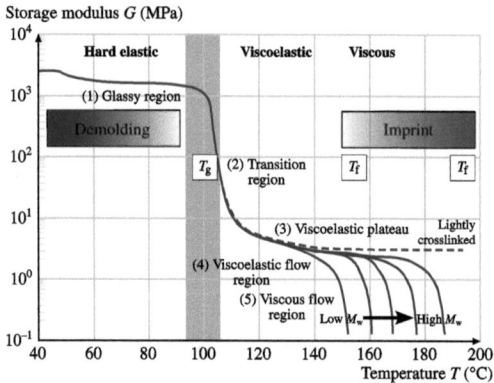

Figure 2.4: Influence of the molecular weight (M_W) and temperature on the storage modulus of a polymer with a glass transition temperature of about 100°C [10].

The reduced viscosity of polymers at higher temperatures is a result of the increasing ability of the polymer chains to move freely. This relation is schematically illustrated in Fig.2.4, where the storage modulus with dependence on the temperature for a polymer with a T_g around 100°C is depicted [10]. In the first region, a thermoplastic material can be described through the glassy state. In this region, the chains of the polymer material are static and the filaments cannot slide along each other, which results in a high storage module and good mechanical properties. Further temperature increments result in a higher mobility of the filaments and a decrement in viscosity. The polymer passes various states before the viscous flow state, which is the region where the polymer can be imprinted, is reached. In the viscoelastic state, the polymer behaves like a liquid while maintaining an elastic component. In this region, the viscosity of the PMMA is still too high for imprinting. It is desired to use a low viscosity material to ease

2. Fundamentals and Methods

cavity filling during the imprint. In Fig.2.4 it is revealed that for materials with a low M_W, a lower viscosity can be obtained at lower temperatures. At the same time, M_W should not be too low, because if T_g is too low, the material cannot be used as a hard resist for pattern transfer [10]. In this work, a PMMA with a M_W of 90 kg/mol was used for imprinting [42], allowing both, pattern transfer into the PMMA at moderate temperatures and the use as etch mask for pattern transfer into the substrate.

2.1.2.2 UV-Curable Materials

There are many different material types that can be used as resists for UV-NIL and other UV based technologies. These can range from UV curable polyurethane or epoxy material [43] to vinyl ether or acrylate based materials [44]. Furthermore, resist materials, adjusted to the needs of the applications, are commercially available [45]. In this work, UV curable materials were used as both the resist in an UV-NIL process and as the material for the fabrication of thick polymer parts in a Soft Lithography process.

In order to fulfill the different requirements of both applications, material systems containing epoxy components, vinyl ether components and an initiator were mixed together in different ratios to adjust the material properties to the needs. A detailed characterization of the materials used in this work can be found in [46, 47]. Vinyl ethers and cycloalophatic diepoxides are among the most reactive monomers polymerized by cationic polymerization. It has been demonstrated that vinyl ethers can be used as a reactive solvent for the photopolymerization of epoxy materials [48]. The cationic polymerization of vinyl ethers, epoxies and mixtures of both components have been widely investigated. Sun et al. [49] have studied the reactivity of different epoxy monomers and found out that a bis - cycloaliphatic epoxide (BCE) has the highest reactivity among the tested monomers. Decker et al. also studied the polymerization of various epoxy monomers in mixtures with acrylate [50] and vinyl ether monomers [51]. For the material system of acrylate and epoxy monomers, the reaction mechanism is different, as the acrylate monomers polymerize by radical polymerization and the epoxy monomers by cationic polymerization. It was observed that the physical properties, like hardness and toughness, could be optimized in mixtures. Unfortunately, the negative properties of radical polymerization like oxygen inhibition and shrinkage were still present. The fabrication of polymer parts with a thickness in the range of mm requires a material with very low shrinkage, otherwise high accuracy of the replication cannot be achieved. In the investigations on the material systems based on vinyl ether and epoxy monomers [51], various monomers from each group were analyzed. The materials were analyzed concerning the degree of polymerization in dependence on the exposure time and viscosity as well as the reaction mechanism of the different mixtures. The most reactive monomers were BCE, diglycidyl ether of bis-phenol A (DGE-BPA)

2.1 Nanoimprint Lithography

as the epoxy monomers and a divinylether of triethylene glycol (DVE-TEG) as the ether component. As initiator, a triaryl sulfonium salt (TAS) was used. Furthermore, it has been demonstrated that vinyl ethers react faster than epoxy materials. The polymerization of the epoxy materials occurs faster in the mixture with the ether monomer than the pure epoxy. Other groups investigated the influence of polyol monomers on the polymerization of epoxy materials [52, 53]. On basis of the results of these groups, in [46], material systems adjusted to the requirements for the replication of polymer parts were developed. The research has shown that the best results could be achieved with mixtures of BCE, DGE-BPA and DVE-TEG. The mixtures of polyol and epoxy monomers are not mechanically stable over time due to the hygroscopic character of the polyol monomers. Therefore, in this work, only mixtures containing epoxy and vinyl ether monomers were used. The process parameters for the different applications will be discussed in chapter 6 and 7.

2.1.3 Comparison

There are some important differences in the process sequence between HE-NIL and UV-NIL. Each method has its specific advantages. In UV-NIL a liquid resist material is patterned at moderate pressure and temperature, which is cross-linked and hardened by curing. In the HE-NIL process, a hard thermoplastic material is first heated up in order to lower its viscosity and then the stamp is pressed into the film using high pressure. From these different process characteristics, different advantages and challenges can be derived. Due to the lower viscosity of the materials used in the UV-NIL process, the cavity filling can occur easier than in the HE-NIL process. The HE-NIL process on the other hand is low-cost since non-transparent stamp materials can be used. The biggest challenge in the HE-NIL process, is the process time. Since the process sequence requires a heating and a cooling cycle, low cycle times are difficult to obtain. In the UV-NIL process it is difficult to obtain large areas with a high resolution. When using hard stamps in UV-NIL, a high resolution can be obtained, comparable to HE-NIL, but a step and repeat process is needed for large areas due to the low pressures. With soft stamps, large areas can be structured in a single step and also undercuts can be replicated [16] but the resolution is typically not that high with flexible stamp materials (sub 100 nm structures are difficult to obtain). Commonly used HE-NIL resists, such as PMMA usually have low dry etching durability. This puts a severe limit on the pattern transfer into the substrate by RIE due to the poor etching selectivity between the PMMA and the substrate material, generally Si. UV curable photoresists provide much higher etching selectivity allowing for the fabrication of structures with much higher aspect ratios [33]. The major characteristics of the process variations of NIL are summarized in Tab.2.1.

Table 2.1: Comparison of HE-NIL and UV-NIL with typical parameters (in modified form from [10])

Type of NIL	HE-NIL	UV-NIL
basic process sequence	1) spin coat thermoplastic film 2) place stamp on film 3) heat until viscous 4) emboss at high pressure 5) cool until solid 6) demold stamp	1) dispense liquid resin 2) parallel alignment of stamp with defined gap 3) imprint at low pressure 4) expose with UV light through stamp and cross link 5) demold stamp
Pressure p	20-100 bar	0-5 bar
Temperature $T_{imprint}$	100-200 °C	20 °C (ambient)
Temperature T_{demold}	20-80 °C	20 °C (ambient)
Resist	solid thermoplastic $T_g \approx 60$-100 °C	liquid, UV curable
Viscosity η	10^3-10^7 Pas	10^{-2}-10^{-3} Pas
Stamp material	Si, SiO$_2$, metals opaque	glass, SiO$_2$ transparent
Stamp area	full wafer, 200 mm diameter	25 x 25 cm^2
Stamp contact	facilitated by bending	planarization layer
Embossing Time	from seconds to minutes	< 1 min (per exposure)
Advantage	low-cost, large-area, equipment and stamps	low viscosity, low pressure, alignment through stamp
Challenge	process time, thermal expansion due to thermal cycle	step and repeat needed for large areas, shrinkage
Development needed	alignment, residual layer homogeneity	material variety
Hybrid approaches	thermoset resists: embossing and curing before demolding	thermoplastic resists: hot molding and UV-curing before demolding
Advantage	low temperature variation cycle: demolding at high temperature possible	solid resist full-wafer single imprint possible

2.1.4 Flow Behaviour of Thin Film

The flow behaviour of thin films during imprint processes is a decisive factor in the calculation of imprint times, the general understanding of the imprint process and for the optimization of the stamp geometry. In a thin polymer film as it is used in NIL, a small vertical displacement results in a large lateral flow of the polymer. The flow behaviour of a liquid thin film during imprint has been widely investigated [23, 54–56]. The flow behaviour and the filling of the stamp cavities depends on the geometry of the stamp, the material properties and the conditions during the imprint. The influence of the stamp geometry can be revealed with a simple model [23]. In Fig.2.5 the embossing of a stamp with line cavities is schematically shown. In this example the lines in the stamp have a length L, a height h_r, a width of s_i and a separation distance of w_i.

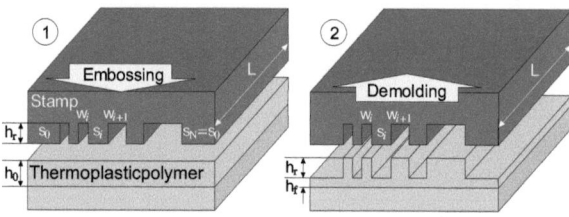

Figure 2.5: Schematic image of a stamp and a coated substrate with geometric definitions before and after the imprint [23].

With the assumption that the polymer is incompressible, the minimal thickness h_f of the imprinted layer after demolding is beneath the lines with the width s_i. This thickness defines the residual layer h_f that has to be removed in order to open the process window for further processing. The minimal thickness of the residual layer can be calculated from the geometrical parameters of the stamp and the initial thickness of the polymer layer h_0 and is given by Eq.2.1, where ν defines the fill factor of the stamp, i.e. the ratio of recessed areas (w_i) to the total stamp area (sum of recessed areas and elevated structures s_i). The correct calculation of the fill factor of a stamp can be difficult for complex stamp geometries.

$$h_f = h_0 - \nu h_r \quad with \quad \nu = \frac{\sum_i w_i}{\sum_i s_i + w_i} \qquad (2.1)$$

This formula only applies for rigid stamps with constant fill factor and the assumption that the polymer flow does not continue toward the borders once the stamp is completely filled. The direct consequence from this simple formula is that the residual layer can be minimized due to the minimization of the elevated structures (s_i) in the stamp, which should be taken into account during stamp design. The minimization of the residual layer is an important factor for further pattern transfer, as this residual layer

2. Fundamentals and Methods

has to be removed in the first step. The second equation (Eq.2.2), the Stefan equation, allows for the calculation of the film thickness h(t) underneath the stamp protrusion [10]. This equation is obtained by solving Navier-Stokes equation with nonslip boundry conditions at the stamp and substrate surface.

$$\frac{1}{h^2(t)} = \frac{1}{h_0^2} + \frac{2F}{\eta_0 L s^3} t \qquad (2.2)$$

From this equation, the embossing time can be calculated if the final thickness of the residual layer is inserted h(t)=h_f.

$$t_f = \frac{\eta_0 s^2}{2p} \left(\frac{1}{h_f^2} - \frac{1}{h_0^2} \right) \qquad (2.3)$$

Eq.2.3 allows the derivation of some important relations between geometric parameters of the stamp and process parameters; these are necessary, in order to obtain good replication results in reasonable imprint times. On one side it can be seen that the pressure p only has a weak influence on the embossing time ($t_f \propto 1/p$), as well as the viscosity ($t_f \propto \eta_0$). However, the viscosity can be significantly changed by varying the temperature [54]. It is obvious that the greatest influence comes from the geometrical parameters of the stamp, i.e. the elevated structures s ($t_f \propto s^2$).

Furthermore, the initial layer thickness affects the form filling process as well. This effect has been observed in experiments [55]. Here a decrease of the initial layer thickness of the polymer was found to give a significant increase in the required embossing temperature to obtain a complete form filling in the same time. A thinner initial film thickness hinders the free polymer flow in the central plane of the film, therefore a higher temperature is needed to decrease the viscosity and obtain a complete form filling in the same time. For a constant temperature (same viscosity), on the other hand, the embossing time must be increased to obtain a complete form filling. This is an important point because the maximum embossing temperature that can be used depends on the available equipment. Therefore, a thinner initial layer thickness cannot always be compensated for completely by increasing the temperature.

Furthermore, the initial layer thickness also affects the residual layer (Eq.2.1). The initial layer thickness has to be chosen adequately in order to find a compromise between low residual layer thickness and short embossing time (at a particular temperature). In Fig.2.6, the filling sequence of a typical polymer is illustrated for a single cavity. Once the stamp has full contact with the polymer (step 4 in Fig.2.6), the flow can only continue to the borders of the stamp [10]. Since the embossing area ($s_i + w_i$) becomes large, as a direct consequence of Eq.2.3 the flow continues much slower. Furthermore,

2.1 Nanoimprint Lithography

the films typically used in NIL are very thin, leading to a high friction at the borders which hinders the polymer flow as well. Therefore, the polymer flow towards the borders is very slow and most unlikely to happen.

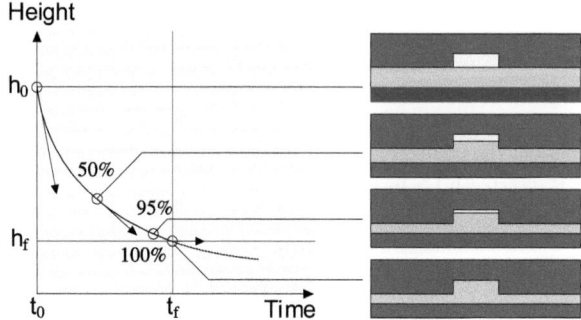

Figure 2.6: Filling sequence of a single cavity during an imprint process [23].

From this simple model some basic rules can be derived. First of all, the stamp geometry can be optimized to improve the flow behaviour of the polymer during the imprint. In order to obtain a thin residual layer after embossing, the filling factor ν should be as big as possible, i.e. the elevated structures (s_i) should be held small (Eq.2.1). Besides the thinner residual layer, the structure size of a single elevated structure should be as small as possible in order to obtain short embossing times (Eq.2.3), as small structure details sink faster than tall ones [23] since less material has to be displaced. Another important factor regarding the flow behaviour is the material which is used for imprinting. The lower the viscosity, the faster the stamp filling occurs (Eq.2.3). In order to achieve a low viscosity at low imprint temperatures and, therefore, low imprint times (due to shorter heating and cooling cycle), a thermoplastic material with a low molecular weight should be used. Besides the material parameters themselves, the thickness of the material also plays an important role as mentioned before. Therefore, a compromise between residual layer thickness and form filling must be chosen when the parameters temperature, layer thickness and embossing time are defined. In the UV-NIL process, the cavity filling can be described with the same equations. However, the low viscosity of the UV curing materials makes the filling process much easier and faster. Therefore, stamp optimization based on this model is more important for the HE-NIL process (i.e. the structuring of thin thermoplastic films on solid substrates).

2.2 Thermoforming

2.2.1 Classic Thermoforming Process

Thermoforming is an established, industrial process for the fabrication of prototype parts as well as for high-volume applications. In this process, a polymer sheet is formed into or onto a mold by applying heat and pressure. The thickness of the sheets varies between a few tens of μm to various mm. The classic Thermoforming process is used for the manufacture of disposable cups and containers, blister packs, trays or other products in the packaging or medical industry. Furthermore, in the auto mobile industry entire vehicle doors or instrument panel skins are fabricated using Thermoforming which makes it a very valuable process for industry [57].

2.2.1.1 Methods of Thermoforming

During the Thermoforming process the polymer sheet can either be molded onto a positive mold or into a cavity of a negative mold as illustrated in Fig.2.7. On the left of Fig.2.7 the concept of a positive molding is shown and on the right side a schematic of negative molding is illustrated. Depending on the chosen concept, the accurate reproduction of surface details is either on the inside (positive molding) or on the outside (negative molding) of the foil.

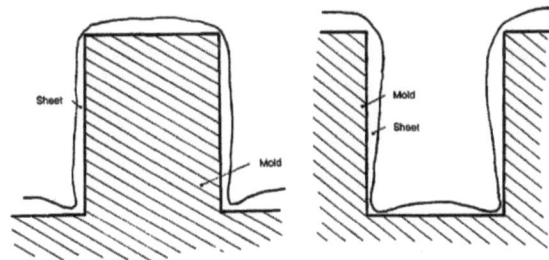

Figure 2.7: Different types of Thermoforming. On the left side the concept of a positive mold is shown and on the right the schematic of negative molding is illustrated [57].

In Thermoforming, different variations of molding a sheet into its final form can be used. In its simplest form, Thermoforming is the stretching of a heated polymer foil into its final shape [57]. During the Thermoforming process, the foil gets stretched until it has contact with the surface of the mold. Once the foil touches the surfaces it stops drawing. A schematic of a typical wall thickness variation during the draw down of a foil into a negative mold is shown in Fig.2.8.

19

2.2 Thermoforming

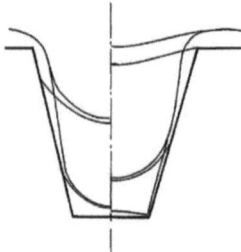

Figure 2.8: Typical thickness variation of a foil during the Thermoforming process[57]

A consequence of this behaviour is that the foil has its maximum thickness at places where it first contacted the mold and its thinnest areas where the mold gets touched last.

2.2.1.2 Foil Heating Mechanism

In the first step after clamping the foil in the Thermoforming machine, the foil is heated to the forming temperature. In general, there are three modes of transferring heat from one object to another which are conduction, convection and radiation.

Conduction describes the heat transfer by solid phase contact. In this work the Thermoforming process is carried out in a tool (see chapter 6.2.1), in which the foil is in direct contact with the tool over a large area. Therefore conduction is the main heat transfer mechanisms in the process as it is done here.

Convection describes the heat transfer in gases or fluids. During the heating step in the Thermoforming process, the foil is in contact with ambient air. Therefore, energy can be transferred from air to the foil when the air temperature differs from the foil temperature. Since the heat transfer during the Thermoforming process is mainly driven by conduction and radiation, convection plays a minor roll as heat transfer mechanism.

Radiation is electromagnetic energy which is emitted from a body because of its temperature. In the Thermoforming process, as it is performed in this work, radiation is the second source of heat transfer. In the heating step, the entire tool is heated and therefore radiation is emitted which heats the foil.

2.2.1.3 Foil Stretching

The foil stretching during the Thermoforming process can be described with the draw ratio. The draw ratio can be calculated in different ways. Two possibilities of describing the draw ratio are the areal draw ratio R_a and the linear draw ratio R_L [57].

Areal Draw Ratio

In this concept, the surface area of the mold is compared with the projected area on the foil. In the case of a cylindrical mold the draw ratio can be calculated as:

$$R_a = \frac{A_{sidewall} + A_{bottom}}{A_0} \qquad (2.4)$$

In this example A_0 is the projected area of the opening of the cylindrical structure to the foil. Examples of more complex geometries can be found in [57].

Linear Draw Ratio

The linear draw ratio R_L is the ratio of the length of a line projected onto the surface of a formed part to its length in the unformed sheet. For asymmetrical parts it is possible to define various linear draw ratios. Normally the largest value for R_L is used. For a cylindrical mold the linear draw ratio can be describes as:

$$R_L = \frac{2s + d}{d} \qquad (2.5)$$

Here, s is the height of a cylindrical hole and d the diameter. For complex geometries it can be easier to use the linear draw ratio, since a projected line on the part surface can be calculated easier than the surface area

2.2.1.4 Material Aspects

Nearly all thermoformable polymers are thermoplastics. The Thermoforming process requires a biaxial stretching of the foil. Even though some thermosetting polymers such as rubbers can be softened when they are heated up above their glass transition temperature, the three dimensional network of rigid thermosetting polymers inhibits the large stretching necessary in the Thermoforming process [57]. Typical thermoplastic material for Thermoforming are polypropylen (PP), polystyrene (PS), polymethylmethacrylate (PMMA) and polycarbonate (PC). Usually Thermoforming requires highly extensible sheets at relatively low stretching loads. The materials can have an amorphous structure but it is also possible to thermoform crystalline materials. A higher degree of crystallinity leads to higher forming pressures. Besides, amorphous polymers are easier to thermoform compared to crystalline thermoplastics, since a higher degree of chain mobility can be obtained. For the different materials, different maximum draw ratios can be obtained. These vary from 4 to 12 for the different polymers [57].

2.2.2 Micro–Thermoforming

Besides the classical Thermoforming process, micro-Thermoforming was developed for the fabrication of low cost microstructured foils [58]. Micro-Thermoforming only differs from the classical Thermoforming process in the dimensions of the structures in the mold and the initial film thickness of the foils. The majority of the applications are in the fields of medical [59], biomedical [60, 61] and fluidic [62, 63] applications. The polymer films used for the Thermoforming process can also be pre-processed. The use of pre-processed polymer films, also called SMART Technology (Substrate Modification And Replication by Thermoforming) was proposed by Giselbrecht et al. [17] . The preprocessing of the substrates includes the use of porous substrates [17], coated substrates [64] or patterned substrates [65–67]. Examples of such structured foils are shown in Fig.2.9. The structure shown in Fig.2.9 (left) was fabricated using Hot Embossing as foil structuring technique, which was then thermoformed into a cylindrical hole with a diameter of 350 µm and a depth of 300 µm. These foils were used as three dimensional structured tissue culture substrates [42]. On the right side of Fig.2.9 a pre-processed foil was molded into a cavity with the same dimensions. For the structuring of this foil X-Ray Lithography was used, allowing for the integration of microstructures with a higher aspect ratio compared to Hot Embossing of about 1-2 [68].

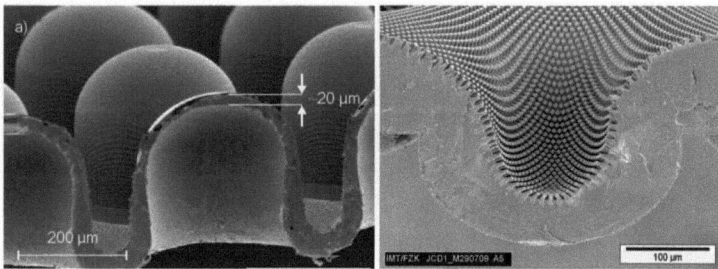

Figure 2.9: Examples of 3D structures obtained using Thermoforming with preprocessed foils[42, 68]

An overview of actualized applications using micro-Thermoforming can be found in [69]. Besides these applications, it has been demonstrated that using pre-patterned thermoplastic foils in the Thermoforming process can be used to integrate a specific functionalization into microstructures. This has been shown for the integration of hydrophobic areas into microfluidic systems [70].

In contrast to the micro-Thermoforming process and applications presented in the above mentioned publications, the Thermoforming process in this work is used for the fabrication of master structures for the Replica Molding process, in order to integrate micro and nanostructured areas into microsystems, so that specific functionalization in polymer devices can be produced. Therefore, the thermoformed 3D structured foil is used as an intermediate structure to fabricate a 3D structured PDMS stamp for the Replica Molding process. A detailed description of the process steps for the fabrication of the master structure will follow in chapter 6.

2.3 Soft Lithography

The Soft Lithography processes were developed in the group of Prof. Whitesides as low cost alternative to classic Photolithography [34]. The term Soft Lithography was used, because all the developed techniques have one common feature, they use a patterned elastomer as mask, stamp or mold [34]. The fabrication and replication of small structures is central to modern science and technology. Further miniaturization poses a challenge to fabrication technologies. Photolithography is still the most successful technology in microfabrication and has been the workhorse of semiconductor industry since its invention in 1959. In order to achieve smaller structure sizes, alternative lithography methods have been proposed including Extreme UV (EUV) Lithography, soft X-Ray Lithography, EBL or Focussed Ion Beam (FIB) Writing. Their development into commercial methods for low cost, high volume processing still requires great ingenuity [71]. The expensive equipment and operating costs of these technologies makes them less accessible for scientists.

Besides these photolithographic methods, non-photolithographic methods like Injection Molding [72, 73], Laser Ablation [74, 75] or NIL [7, 19], capable of the reproduction of sub 100 nm structure details, have been developed. The use of elastomer materials offer some advantages compared to these methods: the fast replication of sub 100 nm structure features [76], structuring over non flat substrates [77] and a great accessibility due to low cost equipment and operating costs. The little capital investment and the fact that they can often be carried out at an ambient laboratory environment make them exceedingly attractive to a wide range of scientists. The most important Soft Lithography methods are Microcontact Printing (μCP)[78], Microtransfer Molding (μTM)[79] and Replica Molding (REM)[43].

In this work, soft stamps are used for both UV-NIL and REM. Since the REM process is used for the replication of 3D structured polymer parts it is described in detail in this chapter. A detailed description of other Soft Lithography methods can be found in [34, 80]. REM processes have been applied to a wide range of applications: the replication of optical devices [9] and fluidical devices [81, 82], the replication of high aspect ratio nanostrucutres [83] and the replication of hydrophobic surface structures [84]. The REM process offers the advantage of replicating complex structures such as 3D structures in a single step. The possible shapes that can be produced with Soft

2.3 Soft Lithography

Lithography methods is beyond those possible with classic photolithographic methods. Furthermore, a high fidelity in the sub 50 nm range can be obtained using this process [76]. Most of the attractive properties of this process are attributed to the material properties of the elastomer stamp used. Unfortunately, there are also some shortcomings and defects which are also caused by the material properties of the elastomer. In the next section, the properties of the stamp material and advantages and disadvantages will be discussed.

2.3.1 PDMS as stamp material for Soft Lithography

PDMS is the most used stamp material for Soft Lithography processes. It is commonly available as two-component silicone rubber material. The cross linking process through a hydrosilylation is done by thermal curing using a platinum catalyst. The reaction equation for the polymerization of PDMS is illustrated in Fig.2.10.

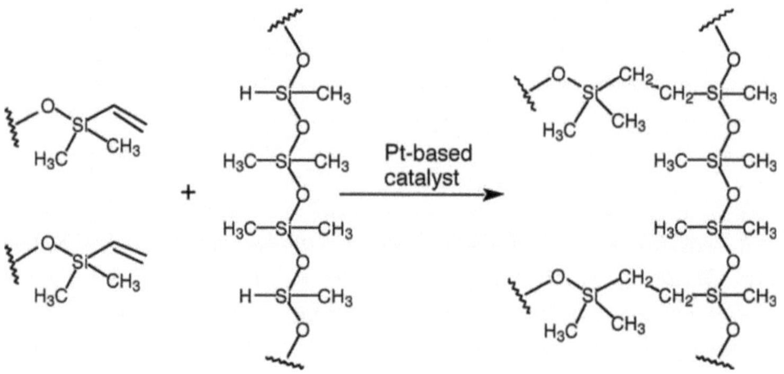

Figure 2.10: Reaction equation for the polymerization of PDMS

In this hydrosilylation reaction, the silyl (SiH) groups of the crosslinker react with the vinyl groups and form the poly(dimethylsiloxane) network as shown in Fig.2.10. A more detailed description of the reaction mechanism can be found in [85]. The uncured PDMS material used in this work has a viscosity of 3.5 Pas and the cured material has a hardness (Shore A) of 45 [86]. The flexibility of PDMS allows a conformal contact of the stamp to the surface of the substrate over a large area [87] and can even compensate for irregularities caused by particles [88, 89]. The elasticity and the low surface energy allows the separation of complex and fragile structures without damages. Furthermore, polymers do not adhere to the surface of the PDMS. Besides these factors, PDMS has good optical properties and is transparent down to wavelengths of 300 nm which makes it an attractive candidate as stamp material for UV based applications. It is a durable elastomer and hundreds of replications can be made from a single stamp [71].

2. Fundamentals and Methods

Unfortunately, the properties of the PDMS, which are responsible for the advantages of Soft Lithography processes, also lead to the technological limits of this method. The limits are summarized in Fig.2.11. It can be seen that the aspect ratio of the structures is a critical factor; if the aspect ratio of the structures is too high (Fig.2.11 left), the structures collapse and pairing occurs. An example of the defect mechanism is also depicted in Fig.2.12. Here a hexagonal structure was replicated in PDMS. The structure has a web width of 100 nm and a depth of 300 nm (Fig.2.12 left). The same effect can be noticed for the line structure where the linewidth is 200 nm and the depth is also 300 nm (Fig.2.12 right). The stability of the PDMS depends not only on the aspect ratio of the structure but also on the form and tone (see section 7.2).

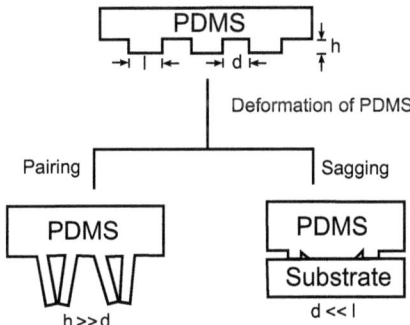

Figure 2.11: Defect mechanism in Soft Lithography methods caused by the mechanical properties of PDMS. The limits are attributed to the geometry of the stamp [34].

Besides the aspect ratio of the structure, the distance between structure details also plays an important role for the stability of the PDMS stamp (Fig.2.11 right). If flat structure details are separated over a large distance, the PDMS can sag in between the structures and the stamp contacts the substrate resulting in a incorrect replication of the stamp.

The stability of a PDMS stamp can be improved by adjusting the mechanical properties of the PDMS to the needs of the application. Schmid et al. [90] have changed the composition of PDMS resulting in a higher cross linking of the material and therefore in a higher hardness of the material. The new PDMS material (h-PDMS) and its higher mechanical stability allows a higher resolution and the previously mentioned problems can be minimized. At the same time, the material maintains its surface properties and the ability to facilitate the separation of the structures.

Another advantage of the PDMS materials is that it can be bonded to PDMS, Si or SiO_2 by plasma treatment [91]. This offers the possibility of fabricating hybrid stamps combining the properties of h-PDMS and normal PDMS [92], as well as the possibility

2.3 Soft Lithography

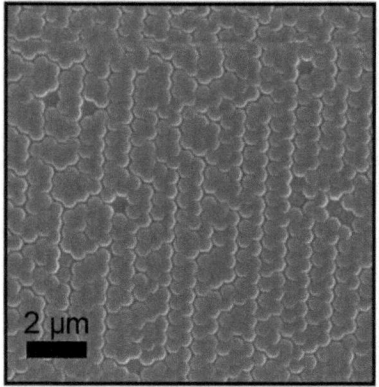 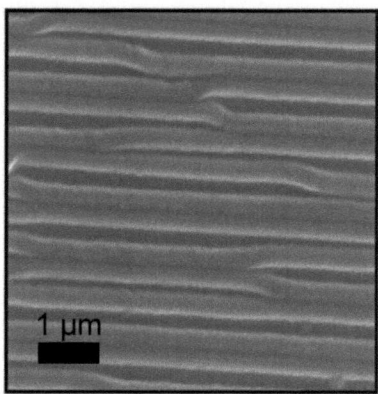

Figure 2.12: SEM pictures of the defect mechanism of PDMS. On the left side a hexagonal structure is shown and on the right side a lines and spaces structure. In both pictures the structures collapse and pairing occurs caused by the aspect ratio of the structures.

of bonding the PDMS on a hard backplane for the fabrication of stamps for UV-NIL or Micro Contact Printing.

Further resolution limitations are caused by the viscosity of the PDMS, which may not be sufficiently low and can cause problems in the form filling of nanostructured areas. The form filling can be improved by using solvents like triethylamine, toluene or hexane in order to lower the viscosity and therefore facilitate the filling of the structures [93]. Furthermore, the wettability of the prepolymer on the substrate surface can be optimized as well by choosing the proper solvent. The optimization of the filling process in combination with the possibility of adjusting the mechanical properties of the PDMS allows for the fabrication of nanostructured stamps with good mechanical properties and a high resolution. These properties are especially interesting for the fabrication of stamps for UV-NIL and Micro Contact Printing, as these applications require high resolutions and the flexibility of the stamp is of minor importance because the replication normally occurs on a flat substrate.

In the case of Replica Molding, the flexibility is more important because the demolding of thicker polymer parts is facilitated by a more flexible stamp. In this work, the focus was the fabrication of 3D structured polymer parts containing high aspect ratio nanostructures. Therefore, flexibility is a key factor for a successful demolding. In order to separate the replicated parts from the PDMS stamp, the mechanical properties of the PDMS were not changed. The fabrication of 3D structured polymer parts has also been reported by other groups. Xia et al. [76] used a stretched, structured PDMS stamp to fabricate 3D structured polymer parts. This method has some drawbacks, because the mechanical deformation of the PDMS stamp leads to a deformation of the

2. Fundamentals and Methods

nanostructures which might change the desired functionality. Furthermore the shape of the microstructures is limited due to the limited ability of stretching the PDMS. Losic et al. [94] replicated micro and nanopatterns from a diatom biosilica by REM. Here a 3D structured polymer part for optical applications was fabricated. This method represents a very common approach as the master structure already exists and no additional work has to be done. This method was also used for the replication of a lotus leaf to fabricate a surface with superhydrophobic surface properties [95, 96]. Unfortunately, all these methods do not allow the integration of functionalities with specific surface properties into a tailored microsystems. In order to do so, new strategies for the master fabrication have to be used to fulfill this goal. In this work, a Thermoforming process was developed for the master fabrication allowing the integration of structured areas into specific areas of a microsystem.

3 Stamp Fabrication

3.1 Stamps for HE-NIL

The fabrication of nanostructured stamps is one of the biggest challenges in Nanoimprint Lithography. The quality of the imprints and of the replications depends highly on the quality of the stamp itself. Conventional lithography methods like UV Lithography reach their limits when it comes to the fabrication of structure details in the sub 100 nm range. Therefore alternative strategies have to be considered in order to fabricate high quality stamps for Nanoimprint Lithography. In this work the stamp fabrication was performed with EBL. Before the fabricated stamps can be used for imprinting, an anti-sticking layer has to be deposited on the stamp in order to easily separate it from the resist after the imprint. In the following sections, the process steps for the fabrication of nanostructured stamps using EBL and the deposition of the antisticking layer will be discussed.

3.1.1 Fabrication Process for Si Stamps

Silicon is one of the most important materials in microtechnology. Reasons for its importance are the good availability, the well explored technology, the possibility of fabricating it in extremely high quality at reasonable prices and the monolithic integration of mechanics and electronics within the same chip. The good mechanical and thermal properties of the material also make it a good candidate for the fabrication of imprint stamps. The Young modulus of the silicon varies between 130 to 185 GPa and it shows neither material fatigue nor plastic deformation. Furthermore, the high melting point of the material of 1415 °C offers a very good thermal stability. It is for these reasons and the availability of fabrication technologies and processes that the imprint stamps used in this work were made from silicon. The structuring of the silicon was performed by EBL in PMMA and standard MEMS (Microelectromechanical systems) processes. An overview over the process steps is illustrated in Fig.3.1

In the first step, a silicon substrate is dehydrated in a vacuum oven at 200°C for at least 15 min before a 150 nm thick PMMA layer is spincoated on top; this is followed by a backing step at 180°C for 20 min in order to remove all solvent in the spincoated layer. The PMMA used as a resist for the EBL has a comparatively high molecular weight of up to 2200 kg/mol. Prior to the EBL, the wafer is cleaved along the silicon crystal direction into chips of a typical size of 2 x 2 cm^2. EBL was performed using a

3.1 Stamps for HE-NIL

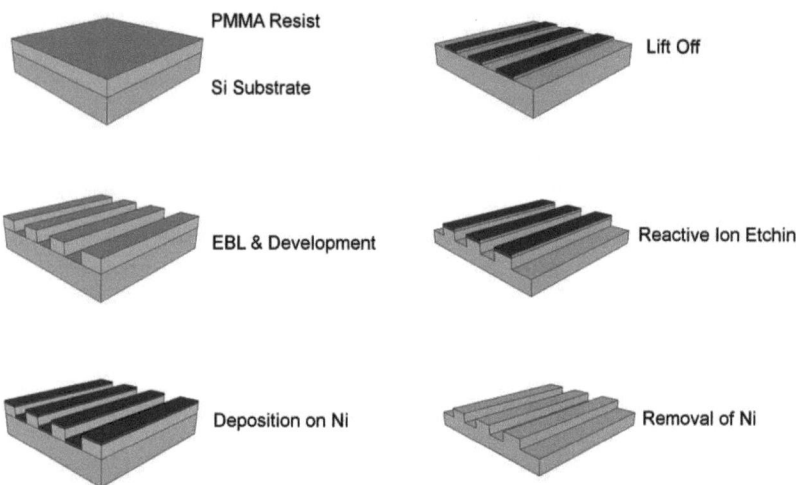

Figure 3.1: Schematic overview over the process steps for the fabrication of imprint stamps.

100 kV electron beam lithography system (EBPG5000, Vistec). This EBL system has a resolution and an overlay accuracy in the sub 10 nm range and a holder for a maximum area of 4 inch is available at the Helmholtz Zentrum Berlin (HZB). After the EBL, the resist was developed for 10 s in AR600-50 developer (Allresist GmbH, Germany) followed by a 10 s rinse in the AR 600-60 stopper, and a 10 s rinse in isopropyl alcohol. Finally, the substrates with the structured PMMA layer were rinsed in de-ionized water and blow dried using pure nitrogen gas. Subsequently a thin Ni layer (approximately 10 nm) is deposited on the substrate by means of thermal evaporation in a high vacuum evaporator A700Q (Leybold GmbH). The thermal evaporation of a Ni wire with a purity of 99,9% (MaTeck GmbH) was performed in a tungsten evaporation boat at about $2*10^{-6}$ mbar with an evaporation rate of about 0.5 to 1 nm/sec. Then a Lift Off process was performed using N,N-dimethylformamid (DMF) to create a metallic mask on the substrate. In contrast to the dissolution of PMMA in aceton, the DMF results in a moisture expansion which favours the Lift Off as the material on the side wall of the structure is lifted as well, resulting in a clean metal mask. For the Lift Off process, the chips were left for several hours in DMF before finishing the process using an ultra-sonic assisted step, to obtain a clean metal mask on the chips. The created Ni mask was subsequently used as an etch mask in the following RIE step. The RIE was performed in the Plasmalab 80 Plus RIE etcher (Oxford Ltd, United Kingdom). The process was performed using SF_6 as etch gas and C_4F_8 as passivation gas to obtain vertical side walls. The gas flow of SF_6 and C_4F_8 was 20 sccm and 15 sccm respectively, and the pressure in the chamber was set to $2*10^{-2}$ mbar. The process involves the use

3. Stamp Fabrication

of an inductive coupled (ICP) and a radio frequency (RF) plasma with a power of 220 W and 20 W respectively. The resulting etchrate of Si in this process is 15 nm/min and a selectivity of over 50 to the Ni mask can be obtained. This allows the use of very thin Ni layers as an etch mask which facilitates the Lift Off process. After the Si etch process the remaining passivation layer is removed using an oxygen plasma with a pressure of $1*10^{-1}$ mbar and a power of 50 W for 2 min. The Ni mask is removed after the etch process using a 25% solution of hydrochloric acid.

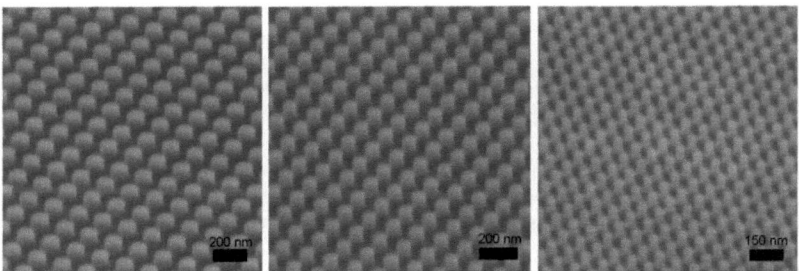

Figure 3.2: Results from the fabrication process for the imprint stamps. The images show various pillar structures with diameters of 100 nm (left), 75 nm (middle) and 50 nm (right).

Results from the stamp fabrication process are illustrated in Fig.3.2. The images show pillar structures with various diameters ranging from 100 nm (left), 75 nm (middle) to 50 nm (right). The structures were etched 100 nm into the substrate. The resulting vertical sidewalls can clearly be noticed.

3.1.2 Anti-sticking Layer for the Stamps

Before the stamps can be used for imprints, the surface of the stamp has to be treated to obtain anti-sticking properties, since an untreated surface results in a destruction of the imprinted pattern [42]. Furthermore, the time-consuming cleaning of the stamps can be avoided due to the use of an antisticking layer [97]. It has been demonstrated that fluorinated silanes can be used to obtain antisticking properties on a Si- or SiO_2- based stamp [97, 98]. In this work (Tridecafluoro-1,1,2,2-tetrahydrooctyl)-trichlorosilane (obtained from ABCR GmbH) short F13-TCS was used as chemical compound to obtain antisticking properties of the Si surface. This molecules bond to the surface of the stamp as illustrated in Fig.3.3.

These amphiphilic molecules are especially suitable for the creation of anti-sticking layers, as they consist of a hydrophilic reactive part and a hydrophobic, non reactive part responsible for the surface properties. The hydrophilic part of the molecule bonds

31

3.2 Stamps for UV-NIL

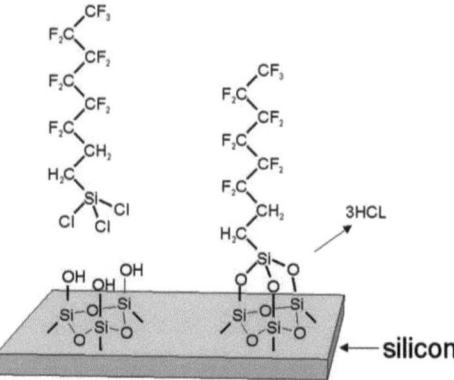

Figure 3.3: The passivation of the surface is obtained using F13-TCS. The molecules consists of a reactive and a non reactive part. The reactive part form covalent bonds with the surface of the stamp and the non reactive hydrophobic part results in a decrease of the surface energy and thus the antisticking properties [42].

covalently to the surface of the stamp while eliminating hydrochloric acid. In this way the formation of a dense layer of the F13-TCS takes place, resulting in new surface properties. The deposition of the silanes occurs from the vapor phase, the method is based on a procedure described in [97]. In order to obtain good bonding properties of the molecules, the surface of the stamp was treated with an oxygen plasma with a RF power of 50 W for 60 s. This removes the organic residue, which was deposited in the previous etching step. The chemical purity of the Si surface plays an important role for the quality of the deposited passivation layer. The influence of the chemical composition of the surface on the quality of the anti sticking properties of the surface has been investigated in [42] using X-Ray photoemission spectroscopy (XPS). The results have shown that the cleaning step of the Si surface results in more uniform deposition of the F13-TCS layer. After the cleaning step, the stamp is put into a petri dish containing a hole in the cap. This assembly is then put into a vacuum oven at 200°C under vacuum for 10 min. Then the oven is purged with nitrogen and 40 μl of F13-TCS is introduced to the petri dish through the hole in the cap. After cooling, the stamp is rinsed in aceton and isopropyl alcohol each for 5 min in an ultrasonic bath to wash off excess F13-TCS. After that the stamps are ready to use for imprint.

3.2 Stamps for UV-NIL

In contrast to HE-NIL, where a thermoplastic resist is mechanically deformed using temperature and pressure, in the UV-NIL process a liquid material is cured using

3. Stamp Fabrication

UV radiation. Therefore, suitable stamps need to fulfill different requirements for this process. Whereas stamps for HE-NIL need thermal and mechanical stability due to the process conditions, the stamps for UV-NIL need good transparency for UV radiation. The resists used for UV-NIL have viscosities which are typically in the range of 10^{-2} to 10^{-3} Pas, the thermoplastic resists used for HE-NIL have typically viscosities of 10^3 to 10^7 Pas at the imprint temperature [10]. Therefore, the UV-NIL process can be performed with less pressure, and the heating step to lower the viscosity of the resist is not necessary. The main difference between suitable stamp materials is that hard stamps offer higher resolutions than soft stamps. In contrast to this, soft stamps facilitate the demolding process due to their flexibility.

In this work, only soft stamps were used for the UV-NIL process. PDMS (Elastosil 601, Wacker) was used as the stamp material, which offers a few advantages compared to hard glass stamps. When using PDMS no additional anti-sticking layer is necessary, as the surface of the PDMS consists of silanes which have anti-sticking properties. This, together with the flexibility of the material (hardness shore A 45 [86]) allows the stamp to demold without problems. Another advantage of UV-NIL is the shorter process time, caused by the faster cavity filling of the liquid resist and the lack of the unnecessary heating and cooling cycle.

The use of PDMS stamps in UV-NIL offers the previously mentioned advantages but has some limitations regarding the resolution as discussed in the previous chapter. In this work, the UV-NIL process was used for the replication of structures in the range of a few hundred nanometres. In order to fabricate a stamp for UV-NIL a master structure has to be fabricated containing the desired structures for replication. The master structures were fabricated using various photolithography processes like Interference Lithography, EBL and conventional Photolithography. For the fabrication of the PDMS stamp the master structure was put into a metal frame. The maximum size of the master structures for the metal frame are 4 inch substrates. A schematic overview of the fabrication process of the PDMS stamp and the following UV-NIL process is illustrated in Fig.3.4. Once a master is fabricated (A) the PDMS prepolymer is mixed according to the manufactures specification[86], degassed for at least 5 min to remove gas bubbles, and cast over the master structure in the metal frame (B). Then a metal plate is brought into contact with the liquid PDMS to define the backside of the stamp. The PDMS is thermally cured on a hot plate for 30 min at 70°C and peeled off the master structure (C). In this way a PDMS stamp containing the negative relief of the master structure is fabricated. For the UV-NIL step the UV curing material is spincoated on the substrate (D) and the PDMS stamp is brought into contact with the liquid resist. Due to the low viscosity of the resist, the cavities in the stamp can be filled within seconds and no pressure has to be applied. Then the resist is cured by UV exposure (E) and after curing, the stamp can easily be demolded from the substrate

3.2 Stamps for UV-NIL

due to the flexibility and the surface properties of the PDMS (F). The process steps and the parameters for stamp fabrication and structure replication are also summarized in the Appendix (process 2 and 4)

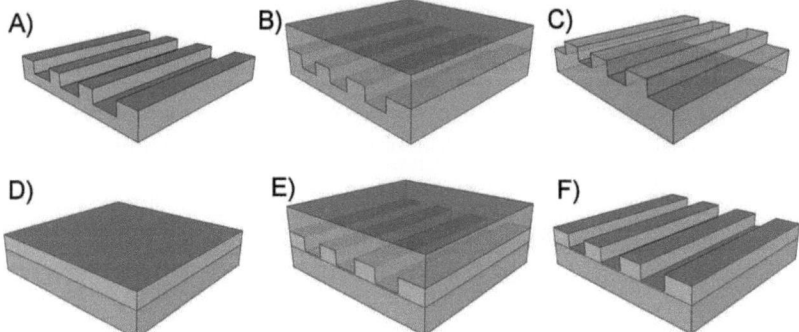

Figure 3.4: Process steps for the fabrication of a PDMS stamp and the replication of a pattern using UV-NIL. In the first step a master structure is fabricated A). The PDMS prepolymer is cast over the master, thermally cured B) and released C). A coated substrate D) and the PDMS stamp are brought into contact E) and the resist is cured by UV exposure. In the last step the stamp is released leaving the structured pattern on the substrate F).

4 Replication of 2D Nanostructures

One goal of this thesis was to develop the HE-NIL process at the Helmholtz Zentrum Berlin (HZB). In this chapter the development of the process steps for the replication of nanostructures are described. In the first section of this chapter, a brief description of the equipment used for the NIL process will be given.
In order to transfer the structures into the substrate two different strategies were used which will be described in detail in section 4.2. The NIL process was further used for structuring of functional layers like metallic materials. In the next chapter, some application developments based on HE-NIL will be described in detail.

4.1 Imprint Machine HEX03

In general an imprint machine consists of a chamber with two parallel, heatable plates. The chamber can be evacuated in order to avoid entrapped air in the imprinted film. The stamp and the coated substrate are introduced into the chamber and then heat and pressure are applied. The system is connected to a computer to create imprint recipes and to monitor the status of the imprint process. The maximum sample size varies from 2 to 8 in. in diameter. In this work the Hot Embossing machine HEX03 (Jenoptik GmbH) was used. The maximum embossing area that can be processed is 150 mm x 150 mm and the maximum force that can be applied is 200 kN with increments of 10 N. The vacuum system can reach a minimal pressure of 1 mbar and the heat and cooling system allows temperatures of up to 220 °C (in order to conserve the sealing rings in the machine a temperature of 190 °C was not exceeded) with a heating and cooling rate of 15 K/min. Therefore, this system fulfills the requirements for nanoimprinting. A more detailed description of this imprint machine can be found in [42, 99]

4.2 HE-NIL Process

The HE-NIL process was performed using a hard stamp made of Si with the method described in section 3.1. As resist, a PMMA granulate (8NL22, Evonik Degussa GmbH) was first dissolved in chlorobenzene and then coated on a Si substrate. The average molecular weight of this PMMA is 90 kg/mol [42], which allows a low viscosity to be obtained at low temperatures (Fig.2.4). This favours complete cavity filling during the process. The usage of a hard stamp required a compliant layer to achieve pressure equilibration [54]. The substrate, the stamp and a 500 μm thick PDMS foil which

4.2 HE-NIL Process

was used as a compliant layer were integrated into the Hot Embossing machine and a polyimide foil was put on top to avoid sticking of the sandwich like assembly to the upper heating plate during the imprint. In the first step of the imprint recipe, a small force of 200 N was applied to stop the assembly from shifting when vacuum is applied. Once a pressure of 10 mbar was reached the plates were heated up to a temperature of 190 °C and the imprint pressure was increased. The optimal pressure during the imprint was found to be about 12.5 MPa. The holding time during the imprint depends on the design of the stamp. At the developing state, the test stamp consisted of various patterns which need different times to obtain a complete form filling. A more detailed discussion on that topic will follow in the next section. In the next step of the imprint, the plates were cooled down, the pressure was released and the assembly was removed from the Hot Embossing machine. The demolding of the stamp and the substrate was done by carefully introducing a razor plate between the stamp and the substrate. After demolding, the remaining residual layer was removed in order to open the process window. To achieve this, an oxygen plasma with an oxygen flow of 10 sccm, a RF power of 20 W and a pressure of 75 mTorr was used [100, 101].

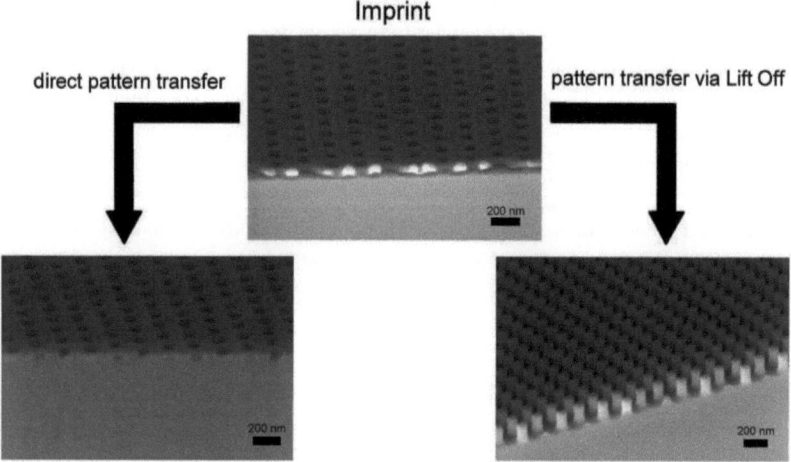

Figure 4.1: SEM images of the results of the HE-NIL process. In the middle, an imprinted structure of a pore field with a pore diameter of 75 nm is depicted. On the left side, the results of the direct pattern transfer is illustrated and on the right, the results of the pattern transfer via Lift Off is shown.

At this point two options arise for transferring the structures into the substrate as discussed before. For the direct structure transfer, the PMMA layer works as an etch mask for the RIE process. The RIE process had to be optimized in order to achieve sufficient selectvity. A selectivity of 2 (etch rate Si to etch rate PMMA) was obtained with an optimized etch process. For the structure transfer via Lift Off a sufficient amount of PMMA must remain on the substrate after the removal of the residual layer, in order to successfully perform a Lift Off. After the removal of the residual layer, a 15 nm thick Ni layer was evaporated on the structured substrate and the Lift Off was performed in N,N-dimethylformamid (DMF). The Ni layer was further used as an etch mask in a RIE process. After etching the structures into the substrate the Ni layer could be removed by a HCl dip. In Fig.4.1, SEM images of a cross sectional area show the result of a test structure containing pillars with a diameter of 75 nm. The imprint was performed with a pressure of 12.5 MPa at a temperature of 190 °C for 10 min. The image in Fig.4.1 (middle) shows the results of the imprinted pillar structure. In Fig.4.1 (left and right) the results of the transferred structures are illustrated (direct patter transfer and pattern transfer via Lift Off respectively).

4.2.1 Influence of Structure Size and Form on Residual Layer

Thin film rheology is one of the most important factors for the understanding of the nanoimprint process, as pattern transfer occurs due to mechanical deformation of the resist. The flow behaviour of thin films affects the homogeneity and thickness of the residual layer. Furthermore, the structured area and the form of the structure have influence on the residual layer (see section 2.1.3).

 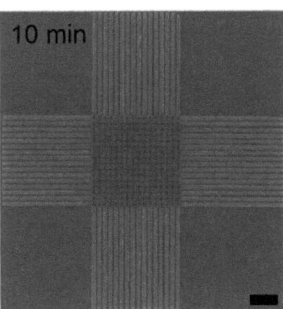

Figure 4.2: Influence of the holding time on the residual layer for a small structure detail. On the left side the holding time was set to 5 min which results in an incomplete structure transfer. A holding time of 10 min led to a complete pattern transfer for this structure detail (right side). The scale bar in both images corresponds to 2 μm.

The homogeneity of the residual layer is very important for the pattern transfer and is influenced by the imprint parameters: temperature, pressure and holding time. A higher temperature, a higher pressure and a longer holding time result in a more complete cavity filling. In this work, the characterization of the formation of the

4.2 HE-NIL Process

residual layer was performed by verifying the Lift Off quality. The maximum imprint temperature, set by the sealing rings in the machine, is 190 °C. The pressure was set to 12,5 MPa and the holding time was varied to characterize the formation of the residual layer. The holding time was varied from 5 min to 25 min and in the following etching process, to remove the residual layer, the etching time was set to 45 s. The residual layer etch process results in an etch rate of the PMMA of about 32 nm/min. The maximum tolerable residual layer thickness in this test is therefore 24 nm to obtain a complete pattern transfer. In Fig.4.2 and Fig.4.3, the results of this test are depicted. The influence of the holding time as well as the dependence on the structure geometry can clearly be seen. In Fig.4.2, a small structured area of the test stamp is revealed. For this structure, a holding time of 10 min is sufficient to obtain a complete structure transfer in the Lift Off process. On the left side in Fig.4.2, the result of the imprint with a 5 min holding time, which was not enough to obtain a homogeneous residual layer, is shown. In the middle of the structure the residual layer was thicker than at the borders, which results in an incomplete removal of the PMMA at the bottom and, therefore, process windows could not be opened completely and the evaporated Ni layer was removed as well during the Lift Off process in this area. In Fig.4.3 a replication of a greater structured area of the test stamp is revealed. In this case the holding time must be increased in order to obtain a complete structure transfer with Lift Off.

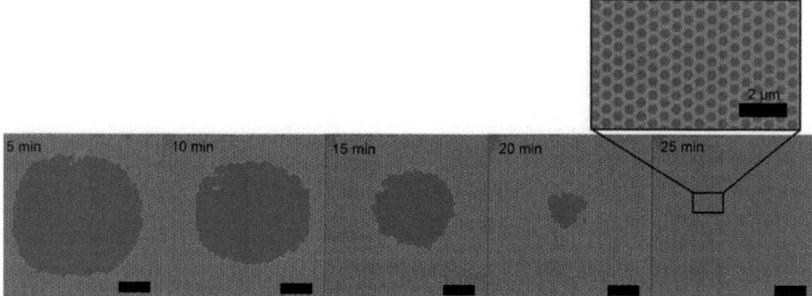

Figure 4.3: Influence of the holding time on the residual layer for a bigger structure detail. Starting at 5 min, the holding time was increased in 5 min intervals until the structure was completely transferred at 25 min.

In section 2.1.3, the dependence of the structure size of a single cavity on the holding time was discussed with a simple model. Here it can be seen that the holding time is not only affected by the structure size of a single element but also by the structured area. The influence of the size of the structured area results from local pressure variations which can not be completely compensated for by the elastic compliance layer [54] resulting in the observed inhomogeneous residual layer.

5 Applications of HE-NIL

In this work, the HE-NIL process was used for structure replication as well as for the structuring of functional layers. In the first section of this chapter, a detailed description of the replication of photonic crystals using HE-NIL will be given. The section includes an introduction to photonic crystals, an overview of the process, a description of the photonic crystal stamp design, and the replication and measurement results.

The structuring of functional materials was used for applications in solar cells and magnetic layers. The developments in these areas are at the initial phases and are still ongoing. The results that are currently available are summarized in the second section of this chapter.

5.1 Replications of Photonic Crystals

5.1.1 Photonic Crystals

Photonic crystals are periodically structured dielectric materials resulting in a periodic modulation of the refractive index which influences the propagation of photons within the material. In short, photonic crystals can be used to create a photonic band gap. The lattice constant of the modulation corresponds to half of the wavelength of the band gap they are designed for. Due to the dimensions of the modulation of the dielectric materials, one dimensional (1D), two dimensional (2D) and three dimensional (3D) photonic crystals can be fabricated. Bragg mirrors, now known as 1D photonic crystals, were first discovered in 1887 [102]. These 1D photonic crystals are widely used as dielectric multilayers in applications like vertical surface emitting lasers (VCSEL) [103, 104]. 3D photonic crystals were first proposed independently by Yablonovitch [2] and John [105] in 1987. Further, 2D photonic crystals have been widely investigated in recent years [106–109]. 2D photonic crystals offer the advantage that they can be used for interesting optical applications like filters [110], splitters [111], sharp bend waveguides [112] and optical resonators [113] for low threshold lasers or single photon sources [114]. Furthermore, the possibility of actively changing the optical properties of the photonic crystals by pattern definition and modulating the physical properties of the materials used, may lead to many novel applications in sensor technology [115]. The most widely used fabrication method for 2D photonic crystals is EBL. This method is a serial fabrication process and is therefore not suitable for mass fabrication. The further development of photonic crystal applications highly depends on a suitable fabrication process for mass production. It has been shown that NIL can be used to quickly fabricate photonic crystals [116, 117]. In [116], a photonic crystal waveguide structure was fabricated in polystyrene with a PBG at a wavelength of 1550 nm. In [117], a photonic crystal structure was fabricated in silicon using NIL and compared with a 3D finite-difference-time-domain (FDTD) simulation. The results have shown good accordance between the fabricated structure and the simulation.

One challenge for the fabrication of photonic crystals is the introduction of microstructured waveguides for in- and out- coupling of the light. Until now, the fabrication of photonic crystals was mainly focused on the fabrication of the nanostructured photonic crystal structure without waveguides. For novel applications, the integration of photonic crystals into waveguides is a decisive factor. Further, the use of adequate materials with good optical properties can improve the performance of the photonic crystals.

In this work, the NIL process was optimized to achieve the structuring of both the photonic crystal structures for applications in the visible range and the microstructured waveguide structures for in- and out- coupling of the light. In order to obtain high quality photonic crystals, Si_3N_4 was used as material. Because of its reasonable refractive index of 2.0 (in the visible range) and its good transmission above 400 nm [115], Si_3N_4 is an interesting material for photonic crystal applications in the visible range. Furthermore, Si_3N_4 is a standard material in MEMS applications and is compatible with microelectronic fabrication processes which is also a decisive factor for the development and incorporation of photonic crystal structures into an optical circuit.

5.1.2 Optical Properties of Different 2D Photonic Crystals

Investigations on the basic physical properties of the photonic crystal structures that were replicated in this work were done in two doctoral thesis [118, 119]. In these works the design was optimized to obtain high quality photonic crystal structures based on silicon nitride. The fabrication of the structures was done using EBL. In this work the design of the stamp is based on the results of these two theses.

5.1.3 Photonic Crystal Design

The photonic crystal design, i.e. the stamp, plays an important role for successful replication with NIL. The rules for the design optimization that can be derived from the model presented in section 2.1.3 were applied to optimize the imprint stamp. The structured PMMA layer should work as etch mask and, therefore, the goal was to obtain a thin residual layer which was achieved by reducing the elevated structure details (Eq.2.1 -> increasing the fill factor). Furthermore the geometric parameters (i.e. the stamp protrusions) have the biggest influence on the embossing time (Eq. 2.3).

The sample design of the photonic crystal structure is depicted in Fig.5.1. Various photonic crystal structures such as cubic filters, filters with defects and resonator structures with a lattice constant of 270 nm were included.

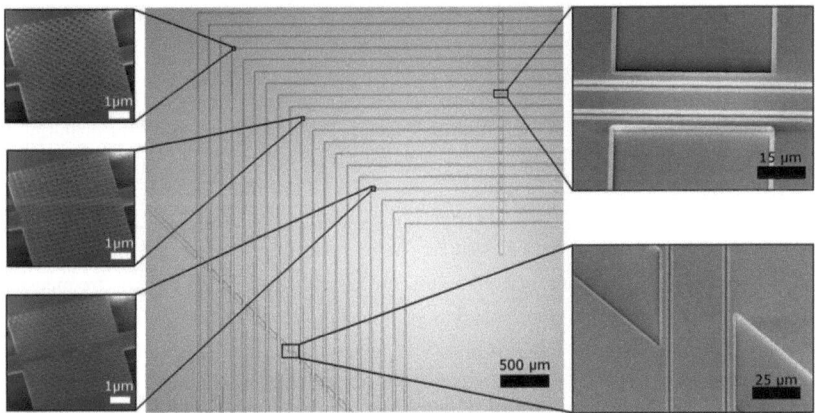

Figure 5.1: Design of the photonic crystal test structure: In the middle of the image an optical microscopy image of the entire, replicated structure is shown. On the left side SEM images of higher magnification of a some photonic crystal samples are depicted. The SEM pictures on the right side reveal the gap for stray light filtering.

In the test structure a filter gap was integrated for stray light filtering. The filter gaps (Fig.5.1 right) are interruptions in the nitride layer to assure that only light which was guided through the waveguide can pass the photonic crystal structure.

The photonic crystals are fabricated as free standing membrane as described in [118]. In this way a high difference in the refractive index between air and the nitride can be obtained leading to a reduction of losses at the interface.

Photonic Crystal Filter Structure

A typical two dimensional photonic crystal consists of a thin dielectric slab perforated with a regular pattern of holes as schematically shown in Fig.5.2. This structure is characterized through its lattice constant (a), the radius of the holes (r) and the thickness of the dielectric slab (t). In such a photonic crystal structure, electromagnetic waves with a specific wavelength can not pass through the photonic crystal allowing the construction of filter structures. The range and the wavelengths in which the photonic crystal works depends on the lattice (e.g. cubic or hexagonal), the period, the

5.1 Replications of Photonic Crystals

radius of the holes, the thickness and the material. Using such structures, photonic band gaps can be produced.

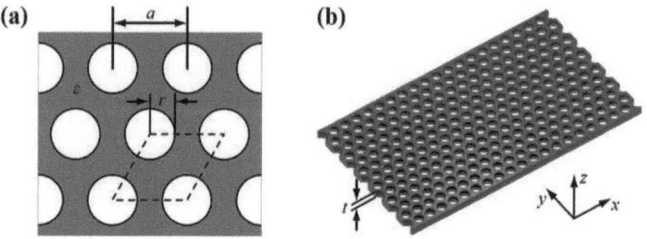

Figure 5.2: Schematic image of a photonic crystal filter structure(b) and the definition of the parameters that influence the function (a)[120]

Photonic Crystal Structure with Point Defect

In the previously mentioned periodic photonic crystal filter structure, the frequencies which lie within the photonic crystal band gap decay exponentially in the photonic crystal latice. However, if the symmetry of such a photonic crystal is broken by introducing a defect into the periodic structure, localized defect modes can occur. The distortion of the lattice includes increased or decreased single holes, shifted holes or just omitting holes completely. Depending whether changes affect small or larger regions, point defect or line defects can be obtained. Due to these defects, different functionalities can be obtained like a photonic waveguide (also W1 waveguide), resonators based on a local change of the effective dielectric constant (double heterostructure cavity), or integrated point defects (L3 cavities). In this work a L3 cavity was integrated into the photonic crystal test design. The design and the effect obtained by such a structure is schematically illustrated in Fig.5.3. In this design 3 consecutive holes in a line are

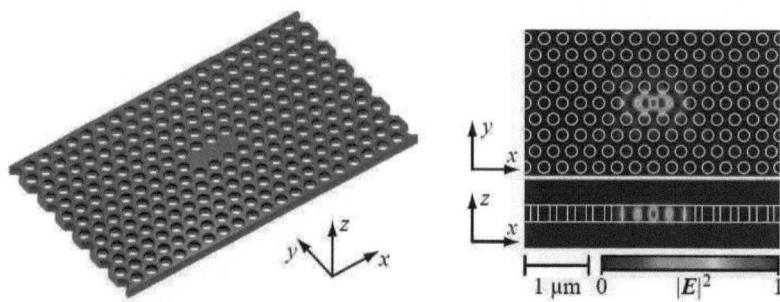

Figure 5.3: Schematic image of a L3 cavity (left) and the simulation result for the L3 cavity (right). In modified form from [120]

5. Applications of HE-NIL

omitted (Fig.5.3a). This local increase in the dielectric constant leads to a defect mode in the photonic band gap (Fig.5.3b). Therefore, a standing wave is formed in the cavity which decays exponentially in all three dimensions. In this way a resonator structure can be manufactured using a photonic crystal structure. A detailed description of the working principle of such a photonic crystal and other effects attributed to defect modes can be found in [120].

SEM images of some of the photonic crystal samples are shown on the left side of Fig.5.1. The samples are connected to waveguides for in- and out-coupling of the light. The spot of the light source was too big to couple it into a single waveguide. Therefore, a bigger area is excited. In order to achieve the excitation of a single waveguide and photonic crystal sample a gap was included to filter out the stray radiation. SEM images of the stray light filter gap are shown on the right side of Fig.5.1. To couple in the light, the waveguide had to be bared. Therefore, the sample was broken perpendicular to the waveguide direction. In this way, the waveguide could be accessed from the side and the light could be coupled in and out. A detailed description of the measurement setup can be found in [115].

5.1.4 Replication of Photonic Crystal with NIL

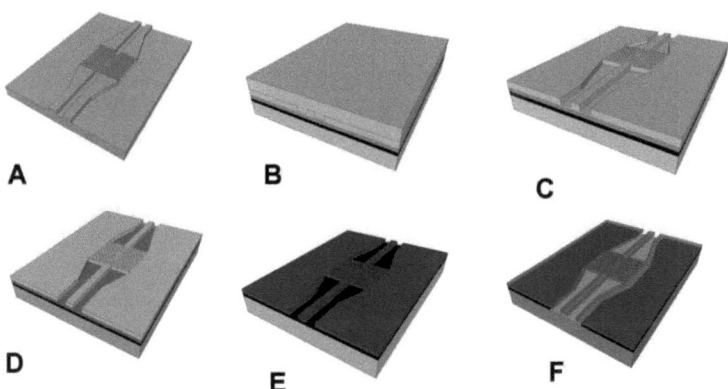

Figure 5.4: Schematic overview over the process steps during NIL: In the first step of the process a stamp is fabricated using EBL (A). This stamp is used for imprinting the structure into a PMMA layer (B). After demoulding (C) the residual layer is etched using RIE (D). Then the structure is transferred into the nitride layer (E) and in the last step the underneath lying oxide layer is partly etched away using HF for obtaining a free standing photonic crystal structure.

43

5.1 Replications of Photonic Crystals

Table 5.1: Process requirements for the replication of photonic crystals using HE-NIL

Physical requirements	Consequence in performance if not fulfilled	Process requirements
Patterned PMMA mask without failures (e.g. caused by contamination on stamp or substrate)	No signal is obtained from the transferred photonic crystal structure or the failures lead to losses and the quality of the photonic crystal decreases	Stamp and substrate must be clean before imprint
Smooth surface of the nitride layer	Intensity losses in the transmission spectra	RIE process must be optimized regarding the surface quality
Photonic crystal structure must be transferred completely	Intensity losses and noise in the transmission spectra	High selectivity in the RIE process high profile of the PMMA mask after the imprint process
Stray light filter gap must be transferred completely	Noise in the transmission spectra caused by stray radiation	
Vertical sidewalls in the photonic crystal structure	Quality loss in the transmission spectra	Anisotropic etch profile in the RIE process

For the fabrication of the photonic crystals a silicon wafer containing a 1.6 μm thick thermally oxidized SiO_2 layer and a 180 nm thick LPCVD Si_3N_4 layer was used. The imprint process was performed in PMMA. In order to optimize the imprint process, various PMMA thicknesses and stamp depths were used ranging from 230 nm to 320 nm and 140 nm to 245 nm respectively. The stamp was imprinted into the PMMA layer at 190°C with a pressure of 12.5 MPa (Fig.5.4 b). The holding time of the imprint was modified in order to obtain the optimal parameters. After demolding (Fig.5.4 c) the residual layer was removed (Fig.5.4 d) and the structured PMMA layer was used as an etch mask for a flourine based RF plasma etching process of the nitride layer. CHF_3 with a flow of 60 sccm was used as etch gas. The RF power was set to 120 W and the pressure to 15 mTorr. After etching the nitride, the remaining PMMA layer was removed by an oxygen plasma (Fig.5.4 e). To achieve a nitride membrane the underlying oxide layer had to be removed. The isotropic etching of the oxide layer was performed using a 50 % solution of pure HF (Fig.5.4 f).

For the replication of photonic crystals, the HE-NIL process has to fulfil some requirements which can be derived from the physical basics. These requirements are summarized in the following Tab. 5.1 .

5. Applications of HE-NIL

5.1.5 Replication Results

In order to fabricate usable photonic crystals with waveguides, the NIL process and the following RIE process had to be optimized. The replication of nano- and microstructures within the same stamp poses a challenge, as the flow behaviour of the PMMA varies for the different structure sizes. The smallest structures on the stamp had dimensions of 80 nm (hole diameter) whereas the biggest structure details were 50 µm (waveguide). Therefore, the amount of PMMA that has to be displaced during the NIL process varies significantly. The PMMA was then used as an etch mask; thus, the final depth of the structures is a decisive factor for complete structure transfer. In the optimized etch process a selectivity of PMMA to Si of 1.8 could be obtained with an anisotropic etch profile. The etching rates for the process were 13 nm/min for PMMA and 24 nm/min for Si_3N_4. The etching times for the different initial PMMA thicknesses and stamp depths were calculated on basis of this data. Further optimization of the etch process regarding the selectivity results in a more isotropic etch profile. Thus the thickness of the final PMMA layer and therefore, the NIL process had to be adjusted to the requirements set by the RIE process. Furthermore, the surface of the PMMA is roughened during the etching process. In order to get a smooth surface of the nitride after etching, the PMMA mask could not be etched away completely to inhibit the transfer of the roughness. To guarantee a good surface quality, the thickness of the PMMA layer should not be thinner than 20 nm after the etch process. To obtain a fully underetched photonic crystal and the removal of the stray radiation at the included gap, the nitride layer has to be etched away completely in this regions to achieve a bared oxide layer. As a consequence from Eq.2.2, small protrusions of the stamp will

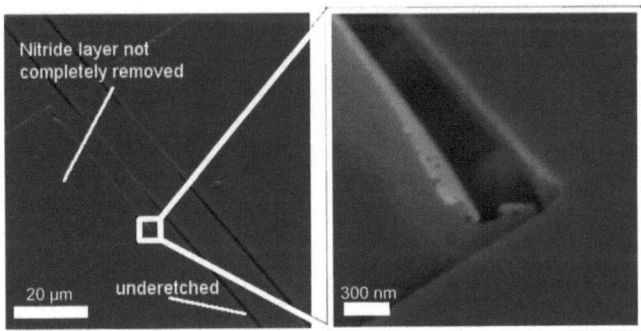

Figure 5.5: SEM image showing the result after etching the sacrificial oxide layer for a sample where the nitride layer has not been etched away completely.

sink faster than wide ones [23] resulting in a distribution of the residual layer in areas with different structure sizes. This results in a thicker residual layer at larger structure details for a specific embossing time. Therefore the largest structures in the design are

45

5.1 Replications of Photonic Crystals

the most critical ones for the pattern transfer, as longer etching times are required to remove the residual layer. In Fig.5.5 the result after etching the sacrificial oxide layer for a sample where the nitride layer has not been etched away completely, attributed to the thicker residual layer in the region of the gap for stray light reduction, is depicted.

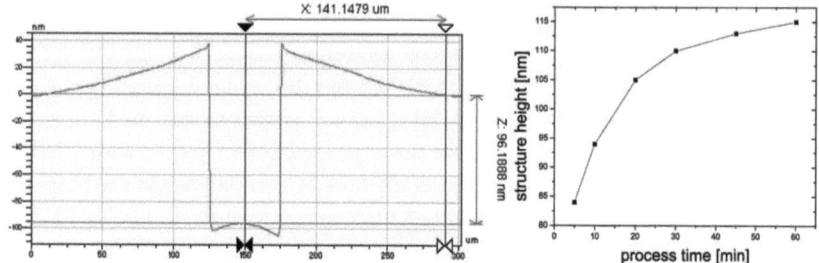

Figure 5.6: Sectional view of an imprinted structure (left side). Structure height in dependence of the process time at constant temperature and pressure (right side).

In order to obtain good replication results, the imprint process was optimized to fulfil the requirements mentioned in Tab.5.1. The most significant points are the thickness of the PMMA layer, the depth of the stamp and the process time. Besides these parameters, imprint parameters like temperature and pressure were analysed as well, but the influence of these parameters were minor compared to the ones previously mentioned. The first problem that occurs when photonic crystals are imprinted into PMMA on a nitride layer is that the PMMA has poor sticking properties on the substrate. Hence the photonic crystal structures were pulled off the substrate during demolding even though the stamp was prepared with an anti-sticking layer. This problem was solved by increasing the PMMA layer thickness. It was found that the PMMA layer should

Table 5.2: The structure height for the optimized imprints for various stamp depths (imprint time 45 min).

Depth of the stamp	PMMA thickness	Effective structure height
140 nm	230 nm	115 nm
170 nm	230 nm	130 nm
215 nm	255 nm	160 nm
245 nm	310 nm	200 nm

be at least 30 nm thicker than the depth of the stamp in order to inhibit tearing off the photonic crystal structures. Furthermore, a higher initial PMMA layer thickness leads to a higher effective structure height in shorter process times (section 2.3.1)[55]. In Fig.5.6 (left) the profile of the imprinted gap for stray radiation reduction in PMMA is depicted. The imprint was performed using a stamp with a depth of 140 nm and the process time was 10 min. It can be seen that a process time of 10 min is not sufficient for obtaining a complete form filling. The effective structure height was 95 nm. In this case, the PMMA did not have enough time to fill the entire structure due to flow issues. In order to optimize cavity filling, the process time was increased resulting in higher structures and structure fidelity (Fig.5.6, right).

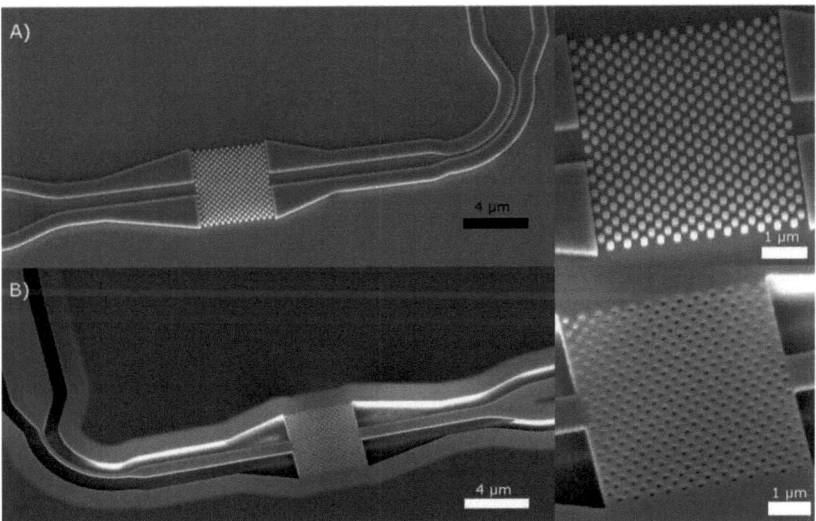

Figure 5.7: SEM picture of photonic crystal structure: In A) a SEM image of the stamp is depicted and on the right side a image of increased magnification of the photonic crystal structure is shown. In B) the results of a replicated photonic crystal structure after etching the oxide layer is illustrated.

This optimization led to an increase of 21% in the effective structure height. Even though the stamp had a depth of 140 nm, the effective structure height that could be achieved after imprinting 60 min was 115 nm. Due to the mentioned requirements the structure height had to be increased in order to obtain a complete structure transfer of the photonic crystal. Therefore, the stamp depth was increased. The effective structure height that could be obtained for different stamp depths and the corresponding PMMA thicknesses are summarized in Tab.5.2. The process time was set to 45 min. The results of the optimized NIL process have shown that the maximum effective height of the micro structures could be increased to 200 nm with a stamp depth of 245 nm and a 310 nm thick PMMA layer. In Fig.5.7, SEM images of the stamp and a replication after HF etching of the oxide layer are shown. Due to the optimization of the NIL and the RIE steps a complete structure transfer of the photonic crystals as well as the critical areas of the stray light filter gap was obtained (Fig.5.1). The images in Fig.5.7 show a photonic crystal and parts of the corresponding waveguide. The structure fidelity lies within the measurement accuracy of the SEM measuring system.

5.2 Outlook

In order to measure the properties of the photonic crystals, waveguides without artefacts had to be produced to obtain measurements from low-loss guiding of the light through the waveguide and the photonic crystal sample. Even minimal artefacts in the replication process would lead to a skewed result in the transmission spectra of the photonic crystal. To characterize the photonic crystal, the waveguide was bared

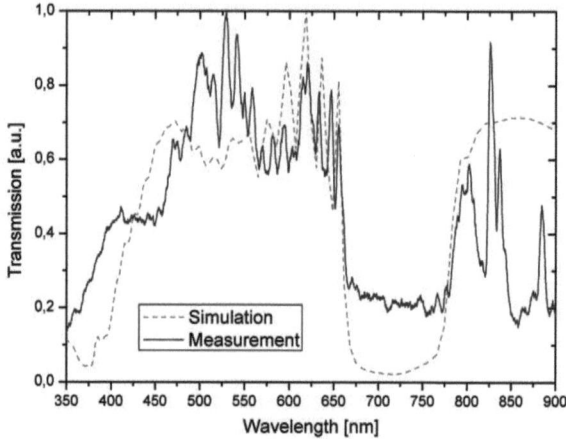

Figure 5.8: Transmission spectra of the photonic crystal structure (continuous line) and the corresponding 3D FDTD simulation (dashed line).

to couple in the light. In Fig.5.8 the results of the transmission spectra for a photonic crystal with a cubic lattice, with a period of 270 nm and a diameter of 130 nm without defects, and the corresponding 3D FDTD simulation (FullWAVE RSOFT LTD.) result is depicted. In the spectra, the PBG at a wavelength of 670 nm to 780 nm can be clearly seen. The size and the form of the measured band gap is in good accordance with the simulation. The result of this investigation has shown that high quality photonic crystals, comparable to those fabricated by EBL were obtained. Therefore, HE-NIL is a suitable process for high volume manufacturing of high quality photonic crystals for applications in the visible range. In order to fabricate optical circuits for more complex applications, further investigations on the basic physical behaviour and possible designs are still needed.

5.2 Outlook

Besides the replication of photonic crystals, the NIL process was also used as a structuring technique for basic investigations on solar cells and the structuring of metallic layers. These investigations are still ongoing; thus, only the basic ideas and the first results of these projects are presented in this chapter.

5.2.1 Structuring of Substrates for Applications in Solar Cells

In order to improve the performance of solar cells, functional materials can be grown onto a Si substrate. The growth of these materials should be structured. In a feasibility study in collaboration with the Institute for Heterogeneous Material Systems from the HZB, it was tested if pre-processed substrates can be used to obtain a structured growth of the functional materials in a metalorganic vapour phase epitaxy (MOVPE) process. As the process parameters in the MOVPE process must be varied, NIL is also a suitable process for the test phase, because many samples must to be prepared to obtain optimal parameters. The pre-processing includes the structuring of <100> as well as <111> Si substrates with various geometries like lines and spaces, pore structures and square holes in different sizes down to 50 nm. The structures in the stamp had a depth of 100 nm and were imprinted into a 110 nm thick PMMA layer. The structured PMMA layer was used as etch mask to transfer the structures 10 - 20 nm deep into the substrate using RIE. After transferring the structures, the PMMA layer was removed by an oxygen plasma. The parameters of the imprint are the same as described in section 4.2. Before starting the MOVPE process, the natural oxide layer on the substrate was removed by HF dip. In the deposition process, no difference between the two crystal orientations was observed. First results of the deposition process are depicted in Fig.5.9. The SEM images show the growth of a CuInSe compound.

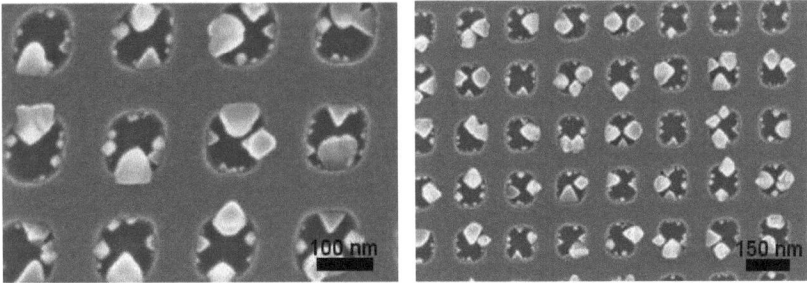

Figure 5.9: SEM images of the first results of the controlled growth of CuInSe compounds on a pre-structured Si substrate.

It is noticeable that the growth of the functional material starts at the edges of the structures. The parameters in the MOVPE process still need improvements to obtain a better control and uniformity of the deposition. These works are currently being performed for other applications and are not part of this work.

5.2 Outlook

5.2.2 Structuring of Metallic Layers

Metallic layers were structured on different substrates to investigate magnetic layers as well as the structuring of catalytic materials for the controlled growth of carbon nano tubes (CNT). The structuring of the metallic layers was performed with a Lift-Off. The presented results were obtained in a collaboration with the Institute of Physics from the Technical University in Chemnitz within the framework of the BMBF project NANETT.

Here, a PMMA layer was structured via NIL and was further used as a mask for the Lift Off Process (see section 4.2). Results from the structuring of magnetic layers on a Si substrate are illustrated in Fig.5.10. Here a 10 nm thick layer stack of Co and Cu was structured. With this process, isolated magnetic islands could be fabricated with dimensions of down to 50 nm.

Figure 5.10: SEM pictures of structured magnetic layers. The structured layer consists of a 10 nm thick layer stack of Cu and Co. The pillars (left side) have a diameter of 75 nm and the lines (right side) a width of 50 nm.

The goal of structuring the magnetic layers is the characterization of the influence of nanostructuring on the basic properties of the magnetic layers as well as the development of new concepts for sensor applications using nanostructuring. In order to obtain information about the basic behaviour of the nanostructured layers, magneto optical kerr effect (MOKE) and superconducting quantum interference device (SQUID) measurements are ongoing. For these measurements various material systems and structure sizes will be characterized. The preparation and measurements are under work. Parallel to the investigations of the fundamental influence of the nanostructuring on the magnetic properties, a concept to use the nanostructuring in an magnetic sensor is under development. This concept makes use of the giant magnetoresistance (GMR) effect. In order to measure this effect, the structured magnetic layer must be electrically contacted. A test structure to measure the resistance of a structured magnetic layer is illustrated in Fig.5.11

5. Applications of HE-NIL

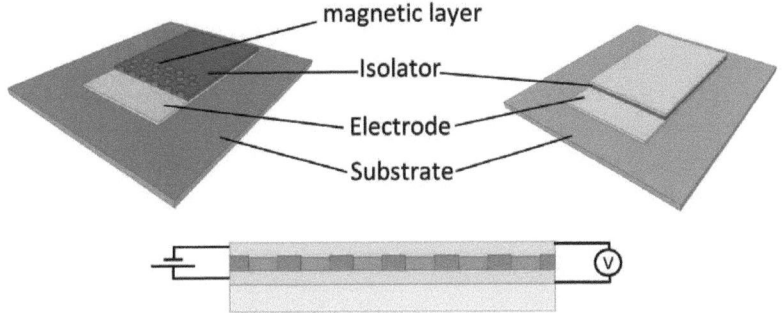

Figure 5.11: Test structure for the measurement of the influence of nanostructuring on the GMR effect. The assembly consist of a substrate, a first electrode, a structured isolator filled with the magnetic material and a second electrode.

The process steps are summarized in Fig.5.12. In the first step an electrode made of Au on a Si substrate is deposited using a sputter process. On top of this layer a Si nitride layer is deposited using a chemical vapour deposition process (PECVD).

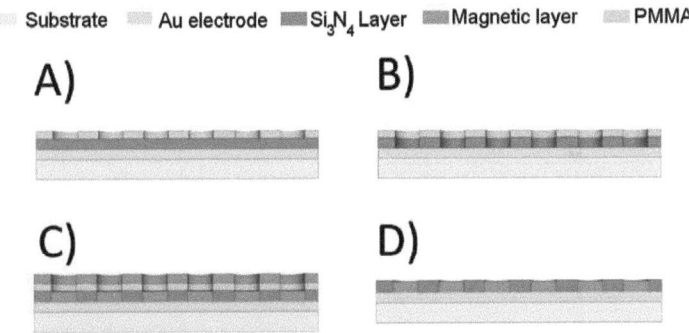

Figure 5.12: Fabricaion process for the test structure to measure the influence of nanostructuring on the GMR effect. In the first step, a PMMA layer is structured on a substrate containing a Au and a nitride layer using NIL A). The structured PMMA layer is first used as an etch mask in a RIE process B). After the deposition of the magnetic layer C), a Lift Off process is performed using the remaining PMMA as resist D).

The magnetic layer is also structured by a Lift Off process by first spincoating a PMMA layer onto the nitride layer which is further structured using NIL (Fig.5.12 A). The structured PMMA layer is first used as an etch mask to bare the Au electrode at the foreseen areas. The PMMA layer is not removed after the etching because the second purpose of the layer is that it works as well as resist for the Lift Off process (Fig.5.12 B). After the etching process, the magnetic material is sputtered into the

51

5.2 Outlook

bared area, where the contact to the lower Au electrode is achieved (Fig.5.12 C). In the next step, the magnetic material on the PMMA layer is removed through the Lift Off process. In this way, the magnetic layer builds isolated islands surrounded by a nitride matrix (Fig.5.12D). The nitride works as isolator to prevent short circuits between the lower and the upper electrode which is sputtered onto the sandwich like assembly. In Fig.5.13 and Fig.5.14, the results of the first feasibility study are depicted. In Fig.5.13 a schematic of the assembly and a SEM picture after the imprint and etching process of the nitride layer is illustrated. The holes in the picture have a diameter of 75 nm and the different layers can clearly be noticed.

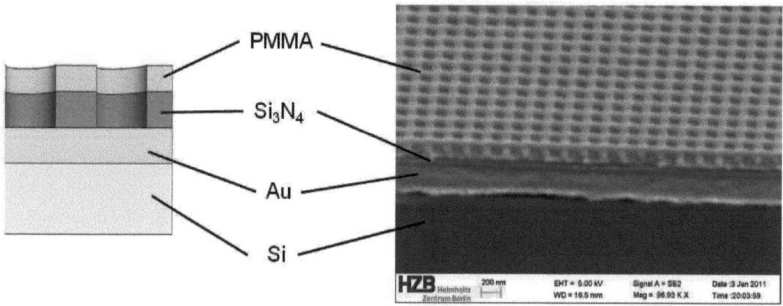

Figure 5.13: Schematic and SEM image of the layer assembly after structuring the PMMA with NIL and etching through the nitride layer.

In Fig.5.14 the same images are shown after the Lift Off process. In the first test, a Ni layer was evaporated onto the structured nitride and PMMA layer, and then, the Lift Off was performed.

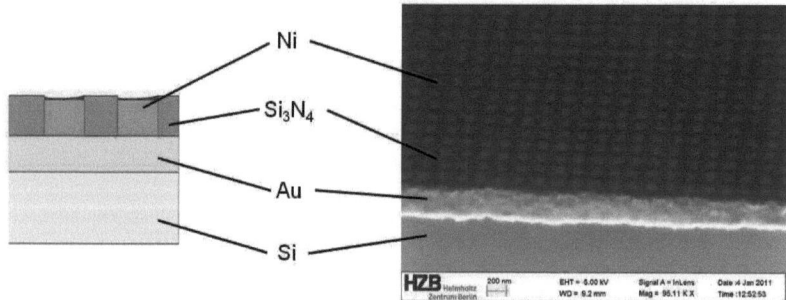

Figure 5.14: Schematic and SEM image of the layer assembly after the Lift Off step.

After filling the nitride layer with Ni which can be seen in the SEM picture, the second Au electrode was sputtered onto the assembly. In order to check whether the concept

5. Applications of HE-NIL

works the electrical resistance was measured using a multimeter. The most critical points in this assembly are contact problems of the Ni (or the magnetic layer) to the lower electrode because of shadowing effects during the evaporation process and the impermeability of the deposited nitride layer. The results have shown that a resistance can be measured between the areas which are connected through Ni. In this first test, many parallel pillar structures were measured leading to a very low resistance (below 1 ohm). The impermeability of the nitride layer was proofed by structuring the upper Au electrode which allowed the measurement of the resistance between the two electrodes separated by the nitride layer. In this areas no resistance could be measured leading to the conclusion that a closed nitride layer was achieved in the CVD process. Therefore, the proof of principle was successful, and design optimizations are currently under work to increase the basic resistance which is necessary to get a strong GMR effect.

6 Fabrication of 3D Structures

6.1 Motivation and Process

In the previous chapters, NIL was used for the replication of 2D nanostructures. In order to fabricate 3D structures out of these 2D ones, a Thermoforming process was used. This is done by covering a microstructured mold insert with a nanostructured foil. The process was applied for the functionalization of microstructures used in fluidic and optical devices. This method offers the possibility of including nanostructured areas into microsystems and, therefore, the fabrication of highly integrated microsystems. As described in section 2.2.2 the Thermoforming process has been used for the integration of microstructures into foil based systems for biomedical applications. In contrast to these works, the Thermoforming process is used as an intermediate lithography step for the fabrication of a PDMS stamp. The direct integration of such 3D structures into a PDMS stamp for the fabrication of full polymer parts is a new approach which allows a specific functionalization of polymer based devices.

In the following chapters, the Thermoforming process and the Replica Molding process, which were used for the fabrication of 3D structured polymer parts, are presented. Besides the development of the process chain for the fabrication of the polymer parts, a new way of foil preparation is also described in section 6.4. Using this improved foil assembly, high aspect ratio nanostructures can be integrated into microstructures. The process chain was used to fabricate a micro fluidic device which was applied to a micro fuel cell as passive fuel delivery system. Besides the possibility of fabricating devices for micro fluidic applications the process was employed to improve optical devices like microlenses. In chapter 8, the development for the fabrication of the previously mentioned application examples is presented.

6.2 Thermoforming: Equipment and Process

6.2.1 Equipment

The fabrication of 3D structured foils using the Thermoforming process required the design of a Thermoforming tool. The resulting structured foil should further serve as a master structure in a Replica Molding process. This point had to be kept in mind when the tool was developed. Besides the Thermoforming tool itself, a pressure system

6.2 Thermoforming: Equipment and Process

had to be designed to control the pressure and the vacuum during the Thermoforming process. The requirements for the tool and the pressure system are summarized in Tab.6.1.

Table 6.1: Requirements for the tool and pressure system for the Thermoforming process.

Process requirement	Implementation in tool
Different mold inserts containing the design of the microstructures should be applied	Integration of an inlay for the mold insert in the Thermoforming tool
The foil must be clamped to be structured using the pressure difference	The foil clamping must be performed between the pressure connection
The part of the tool below the foil must be evacuated to obtain a high molding accuracy	A vacuum connection must be integrated into the Thermoforming tool and the vacuum must be controllable by pressure system.
For the foil structuring a pressure must be applied above the foil	The pressure must be controllable by the pressure system and it has to be assured that the pressure and vacuum part are separated.
A PDMS stamp should be fabricated from the thermoformed foil.	A ring must be integrated into the Thermoforming tool to cast the PDMS prepolymer into and to fabricate the PDMS stamp.

A schematic image of the pressure system is illustrated in Fig.6.1 . The pressure system consists of two sectors, the pressure control (orange frame) and the vacuum control (green frame). The exits of each section are connected to a pressure and a vacuum port at the Thermoforming tool (Fig.6.2, 1 and 3). Nitrogen gas was used as the pressure supply and the pressure was controlled by a pressure regulator unit. This unit allows the regulation of the pressure between 0 and 100 bar. A scroll compressor was used as the vacuum pump and the integrated switches and valves allow for control over the vacuum and the pressure during the Thermoforming process. In this way, the pressure port of the Thermoforming tool could be switched from a vacuum state to a pressure state, while at the normal vacuum port only vacuum and ambient pressure could be charged.

The Thermoforming tool (Fig.6.2) consist of three parts, an upper part (1) with a pressure port, a ring in the middle (2) which is needed for the further fabrication of a PDMS stamp and a lower part (3) with a chuck for the mold insert (6) and the vacuum port. For the Thermoforming process the mold insert is placed in the lower part of the Thermoforming tool, a structured thermoplastic foil (5) is placed on top of the mold insert and the ring and the upper part are put on top to obtain a closed assembly. For the Thermoforming process the assembly is mounted into the Hot Embossing machine and a force is applied for better thermal contact. In the next step the whole assembly is heated up above the glass transition temperature of the foil. The heat transfer from

6. Fabrication of 3D Structures

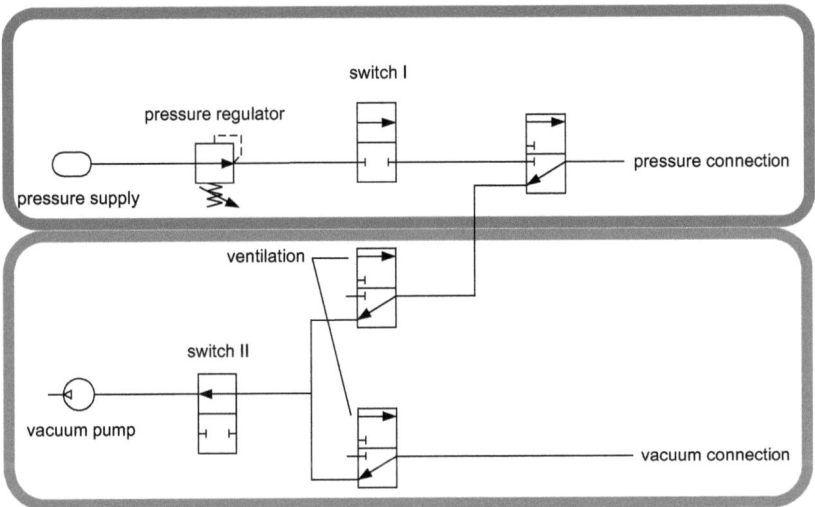

Figure 6.1: Schematic overview over the pressure system used for the Thermoforming process. It consists of two sectors, one for pressure control (orange frame) and one for vacuum control (green frame)

the tool to the foil occurs as described in section 2.2. During the heating process the assembly is evacuated to obtain a good molding quality. Once the assembly has reached a temperature above the glass transition temperature of the foil, the valve at the pressure port is switched and connected to the pressure system. After increasing the system's pressure, the foil covers the mold insert; this results in a 3 D structured foil due to the pressure difference between the upper and the lower part of the tool. In the last step, the assembly is cooled down, it is vented, and the foil maintains its structured shape.

6.2.2 Thermoforming Results

In the first tests, unstructured foils were used to find the correct parameters for the Thermoforming process. As a substrate, a polystyrene foil (provided by Norflex GmbH, Germany) was used. This foil type consists of a blend of polystyrene and a styrene-butadiene copolymer (elastomeric component) and has proven its stability during the Thermoforming process [121]. For testing the device, a mold insert containing a channel structure was used. Various foil substrates with different thickness were available. The minimal foil thickness is 20 μm and the maximum foil thickness is 100 μm. The parameters temperature and pressure were varied to find good values for each foil. The tests have shown that except for the 20 μm foil the temperature can be varied

6.2 Thermoforming: Equipment and Process

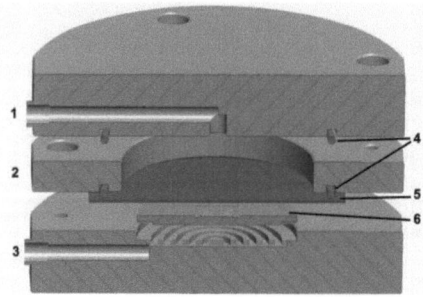

Figure 6.2: On the right side a schematic drawing of the Thermoforming tool is illustrated, the assembly consists of: 1) the upper part with pressure connection, 2) the ring for the fabrication of a PDMS stamp, 3) the lower part with vacuum connection and chuck for mold insert, 4) sealing rings, 5) a structured thermoplastic foil and a 6) mold insert. On the right side a micro milled brass mold insert is shown.

between 100°C and 110°C and pressures of up to 40 bars without damaging the foil can be applied. For the 20 μm foil the parameters, especially the temperature had to be chosen more carefully. The 20 μm foil has shown an increasing tendency to tear at temperatures above 105°C which inhibits a correct molding result. Since it is desirable to use thin substrates to obtain a high accuracy of the structured foil, the best compromise between stability of the foil and accuracy of the final structure was the 50 μm thick foil.

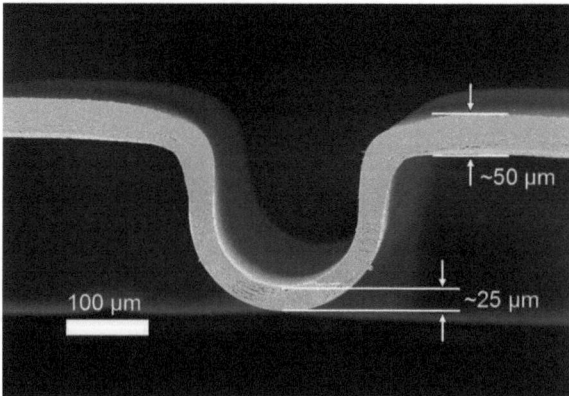

Figure 6.3: Result of the Thermoforming process for an unstructures foil.

A results of the Thermoforming process with an unstructured foil is depicted in Fig.6.3; it clearly shows the thinning of the foil. The temperature was set to 105°C and a pressure of 30 bars was used in the Thermoforming process. The initial foil thickness was 50 μm. After the Thermoforming process a minimal foil thickness of 25 μm could be measured at the thinnest point. The foil was formed over a channel with a width of 250 μm and a depth of 330 μm. The foil was molded about 210 μm deep into the channel. Therefore, the channel was not formed out completely with the applied parameters. For this foil a linear stretch ratio R_L (Eq.2.4) of 2.5 could be obtained (for the calculation, the outer foil length was used). This is 68,7% of the theoretical possible value of 3.64. A higher molding accuracy is obtained with the improved foil assembly which is described in section 6.5.1.

During these initial test it was observed that high pressures and temperatures damage the foil at the sharp borders of the micromilled mold insert. Therefore the mold insert had to be electropolished prior to using it in the Thermoforming process. The electropolishing step is especially important for complex microstructures with sharp edges.

6.3 Foil Structuring using Hot Embossing

6.3.1 Process

In the first attempt, the foil structuring was done by applying standard Hot Embossing to the foil. The process steps for the fabrication of a 3D structured foil is illustrated in Fig.6.4. The imprint process of the foil was performed in the HEX03 with a temperature of 115 °C (the PS foil has a glass transition temperature of 80 °C) and a pressure of 7.5 MPa. (Fig.6.4A)). The structured foil was mounted together with the mold insert (Fig.6.4B)) into the Thermoforming tool. The assembly was put into the Hot Embossing machine, where a force of 15 kN was applied. The assembly was heated up to 105 °C and the molding pressure was set to 30 bars.

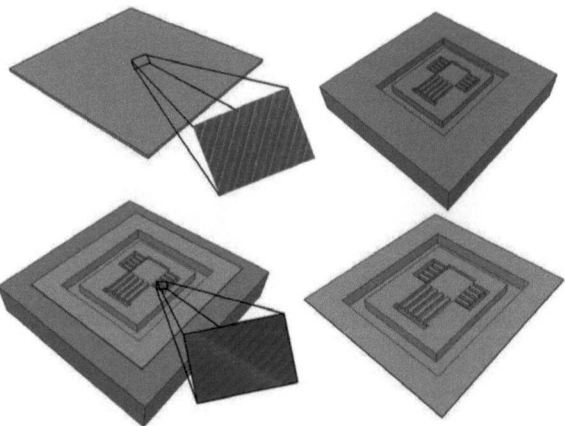

Figure 6.4: 3D structuring of a thermoplastic foil. The thermoplastic foil is micro or nanostructured by Hot Embossing. For the Thermoforming process a mold insert fabricated by micro milling is used. The 3D structuring of the foil is performed by applying temperature and pressure in the Thermoforming tool. After cooling down and venting the assembly, the 3D structured foil can be seperated from the mold insert.

A comparison between the process parameters during the Hot Embossing process and the Thermoforming process is illustrated in Fig.6.5. Both processes begin with a heating step (1,5). Whereas the Hot Embossing process requires a temperature ($T_{imprint}$) well above the glass transition temperature of the thermoplastic material before applying the pressure (2), the Thermoforming process requires lower temperatures (T_{thermo}) when pressure is applied (6). Another difference in the sequence is that in the Hot Embossing process the temperature is held until the cavities in the stamp are filled (2-3). In the Thermoforming process, the molding can be done in the viscoelastic state of the material, therefore, the cooling cycle starts directly after applying the pressure, so no holding time was applied in this process. For both processes the pressure is held until the demolding temperature (T_{demold}) is reached (4,7).

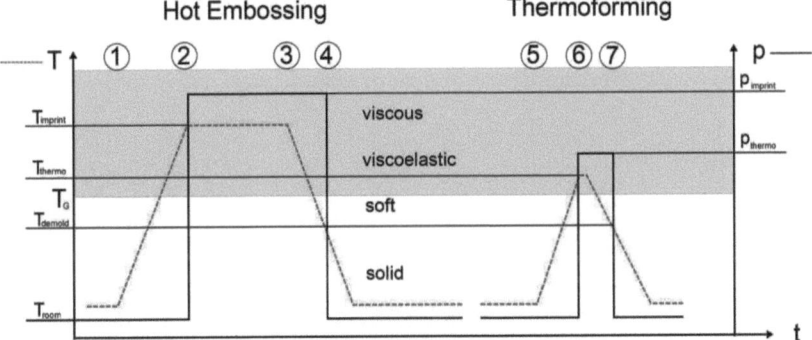

Figure 6.5: Comparison of the temperature pressure sequence during Hot Embossing (left) and Thermoforming (right)

6.3.2 Thermoforming Results using Hot Embossing for Foil Structuring

This fabrication method for 3D structured foils uses different molding conditions in the different process steps. For the imprint of submicrometer or nanometer structures, high temperature and high pressure is necessary for a complete form filling. The Thermoforming process can be performed using less temperature and pressure. Therefore, it is possible to fabricate 3D structures with this advanced lithography method. To achieve a good quality of submicro- and microstructures, the parameters of each step must be carefully optimized. Using a high temperature during the Thermoforming process, leads to a good quality of the microstructures but poor quality of the submicrometer structures. The same effect can be observed if an excessive pressure is used. The optimal parameters for good micro and submicrometer replication in the Thermoforming process were obtained using a temperature of 105°C and a pressure of 30 bar. Process parameters further depend on the design of the microstructures. In the mold insert used here, the maximum depth of the structures is 600 μm, and the general design consists of 10 channels with a width of 200 μm and a depth of 300 μm and 10 channels with a width of 250 μm and a depth of 330 μm. Therefore, a large deformation of the foil has to take place at the channel structures for a high structure fidelity.

Fig. 6.6 shows some examples of structured foils. Fig. 6.6A-D present SEM images of a structured foil. The lines and spaces have a width of 1.6 μm and a depth of 1 μm. The foil was formed over the mold insert containing the previously mentioned channel

6.3 Foil Structuring using Hot Embossing

structures. The edges, seen in the image, are rounded because of an electropolishing step applied to the mold insert. In the images it can also be seen, that lines and spaces are replicated with high fidelity. Using this structuring technique, it is also possible to fabricate undercuts, which cannot be actualized with standard lithography processes.

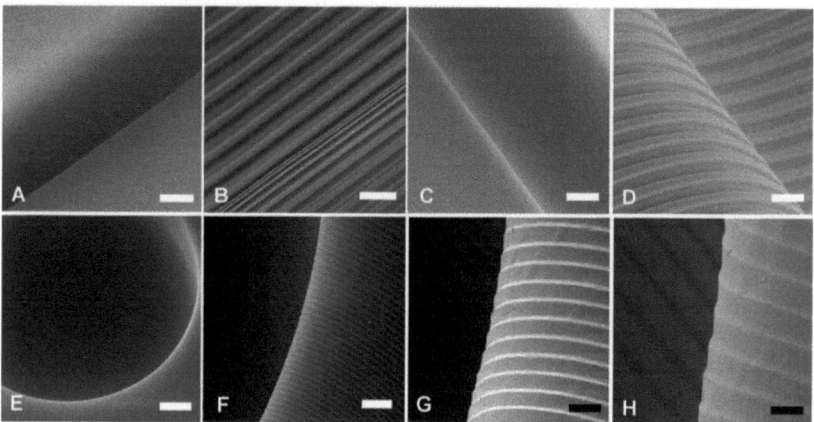

Figure 6.6: Replication results from the Thermoforming process. All pictures show structured PS foils: A, B) sight on a 200 μm wide channel structured with 1.6 μm lines and spaces in channel direction with increasing magnification(scale bar = 60 μm and 6 μm respectively) . C, D)sight on a channel with the same lines and spaces in orthogonal direction with increasing magnification(scale bar = 40 μm and 4 μm respectively). E-H) SEM images of increasing magnification of a blazed grating with a period of 1 μm and a blazed angle of 15° formed over a 600 μm wide, 600 μm deep hole (scale bar = 60 μm, 10 μm,3 μm and 2 μm respectively)

In Fig. 6.6 E-H SEM images of increasing magnification of a imprinted blazed grating which was formed into a 600 μm wide, 600 μm deep hole are illustrated. The grating has a period of 1 μm and a blazed angle of 15°. Fig. 6.6 E shows a overview of the entire pit and 6.6 F-H show the magnification of the edge of the pit. It can be seen, that the profile of the blazed grating maintains good quality after the Thermoforming process. Consequently, this process can be applied for the fabrication of 3D blazed gratings.

6.3.3 Extending the Boundary Conditions

Due to the use of Hot Embossing for the foil structuring, the Thermoforming process has some limitations regarding the molding quality, as the temperature and pressure have to be chosen carefully in order to avoid the destruction of the nanostructures. Therefore, it is difficult to fabricate 3D structured foils containing high aspect ratios.

6. Fabrication of 3D Structures

This has two main reason, first of all it is difficult to fabricate high aspect ratios in these dimensions using Hot Embossing. Even though it was demonstrated that high aspect ratio structures of up to 19 can be obtained with Hot Embossing [122, 123], there are some limitations regarding size, shape, area and materials which makes the fabrication of such structures on a thermoplastic foil quite complicated. The second and more important point is that the thermoplastic foil is heated up above its glass transition temperature and pressure is applied in order to cover the structured mold insert during the Thermoforming process. This is a more significant problem, as the high aspect ratio nanostructure are deformed when temperature and pressure is applied. The parameters limited by the nanostructure quality, also affect the molding quality of the microstructures. Due to these limitations the structures in the mold insert can not be formed out completely. To overcome these limitations a new approach was developed using a different foil assembly. In order to broaden the parameter range in the Thermoforming process, the nanostructuring of the foil is done using the UV-NIL process. This offers two new possibilities. First, the nanostructures can be made out of another material with a higher temperature stability allowing the usage of higher temperatures without affecting the quality, and second, with UV-NIL and the usage of flexible PDMS stamps higher aspect ratios can be obtained. These advantages are attributed to the low viscosity of the epoxy materials which ease the cavity filling and the elastic properties of the PDMS stamp which facilitate the demolding of the structures. The second problem that has been observed, is the damage of the foil at sharp corners. In order to avoid this problem a thin PDMS membrane is deposited on the backside of the foil. The PDMS membrane works as a compliance layer and can compensate the stress impact on the foil at sharp corners. The process steps that are used in this new approach and the results that are obtained are described in the following section.

6.4 Foil Structuring using UV-NIL

6.4.1 Process

As mentioned before, the Thermoforming process can be applied for the integration of functional surface structures into microdevices. This includes the integration of hydrophobic areas into fluidic devices. For the fabrication of hydrophobic surfaces, pillar structures have proven to show good results [124–127], besides the shape of the structure, the aspect ratio also plays an important role for the fabrication of hydrophobic surfaces [128]. A higher aspect ratio leads to a higher contact angle and, therefore, to a larger hydrophobic effect of the surface. To obtain pillar structures with higher aspect ratios, a new approach was developed. In this new approach, the foil structuring is performed by UV-NIL. This has the advantage, that the structured layer can be fabricated in a different material with a higher temperature stability than the thermoplastic foil; this results in a higher stability of the nanostructures during the Thermoforming process.

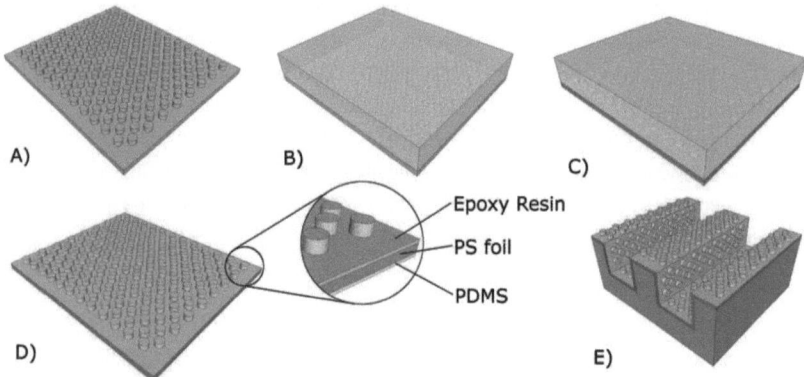

Figure 6.7: Process overview for the fabrication of 3D structured foils using UV-NIL. In the first step a nanostructured master is fabricated A). The PDMS stamp is fabricated by casting the prepolymer over the structured master followed by thermal curing B). The PDMS stamp and the foil coated with a epoxy layer are brought into contact and the epoxy layer is cured by UV exposure C) to obtain a nanostructured foil. In D) the nanostructured foil is depicted. The foil consists of the structured epoxy layer, the thermoplastic PS foil in the middle and the thin PDMS layer on the backside. The 3D structuring is achieved by a Thermoforming step, where pressure and temperature are used to cover a mold insert with the structured foil E).

The fabrication process involves 3 main steps: (1) fabrication of a PDMS stamp from a master structure for UV-NIL (A/B), (2) structuring the foil by UV-NIL (C/D), (3)

6. Fabrication of 3D Structures

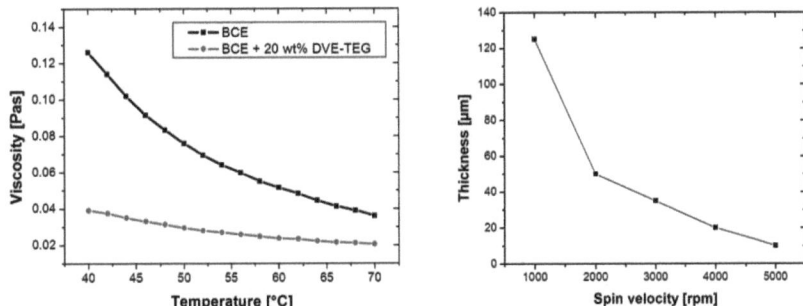

Figure 6.8: Temperature-dependent viscosity of BCE and a blend of BCE with 20% of DVE-TEG[47](left) and thickness of PDMS membranes at different spinning velocities (right)

covering a micromilled mold insert with the foil using Thermoforming (E). The fabrication of the mold insert can also be performed using other technologies like lithography processes, LIGA or EDM. The process steps are described in detail in the following paragraphs and are summarized in Fig.6.7.

In order to transfer a functional structure onto a thermoplastic foil using the UV-NIL process, a master containing the structures that should be integrated into the microsystem must be produced. The initial structuring techniques depends on the functionalization and can be done using e.g. UV-Lithography, EBL or Interference Lithography. As a test structure, a high aspect ratio submicrometer structure is used. The design of the Si master contains pillar structures with various diameters ranging from 600 nm to 6.4 μm. The structure height was 7 μm which resulted in different aspect ratios. For the fabrication of the initial master, UV-Lithography was used for pattern definition which was subsequently tranferred into the substrate using RIE.

From this master structure, a PDMS stamp was produced as described in section 3.2. In this way, the negative tone of the original master structure can be obtained in the PDMS stamp. Prior to the pattern transfer a thin PDMS membrane is spincoated onto the backside of a 50 μm thick PS foil for stress compensation in the following Thermoforming process. As the PDMS prepolymer has a high viscosity, the spinning velocity had to be very high to achieve a thin membrane. In Fig.6.8 (right), the resulting thickness of the PDMS membrane at different spinning velocities is plotted. The PDMS membranes were coated with a velocity of 5000 rpm, which results in a thickness of 10 μm. Additionally, a 10 μm thick layer of an epoxy mixture was spincoated on top of the foil. The mixture contained a bis-cycloaliphatic epoxide (BCE), a divinylether of triethylene glycol (DVE-TEG) and a triaryl sulfonium salt (TAS) as initiator.

In order to obtain adequate thickness of the epoxy layer, the proportions of the com-

65

ponents can be changed as well as the speed during spincoating. The dependence of the viscosity on the proportions of the components and the temperature is depicted in Fig.6.8 (left). A drastic decrease of the viscosity can be noticed in the blend compared to the pure BCE. The possibility of tuning the viscosity allows for the adjustment of the final thickness of the epoxy layer within a wide range. In the next step the structured PDMS stamp was brought into contact with the epoxy layer on the foil. Due to the low viscosity of the epoxy layer, the liquid filled the cavities of the PDMS stamp within seconds. The exposure was performed with a dose of 400 $\frac{mJ}{cm^2}$. After the exposure, the PDMS stamp could easily be peeled off the structured foil.

This structured foil is further used as a substrate in the Thermoforming process as described in the previous section. The structured foil was put into the assembly with the PDMS coated side on the mold insert. Due to the PDMS membrane, the electropolishing step which was necessary before is not needed anymore because the membrane inhibits the destruction of the foil at sharp corners. Here a temperature of 115° and a pressure of 45 bar could be applied which leads to a better molding quality of the microstructures without risking damage to the foil. The nanostructures can also be obtained in better quality because of the higher temperature stability of the epoxy material. The lower viscosity of the epoxy material also helps to achieve higher aspect ratios in the UV-NIL process compared to Hot Embossing as the cavity filling occurs much easier.

6.4.2 Thermoforming Results using UV-NIL for Foil Structuring

This new approach was developed to optimize two characteristics of the process. First, a high accuracy of the microstructures, and, second, avoiding the deformation of the integrated surface structures. The first point can be achieved by applying the PDMS membrane to the backside of the foil as mentioned before. The second point can be obtained by using the UV-NIL process as initial foil structuring technique. In order to transfer high aspect ratio structures onto the foil, the structuring is performed with a flexible PDMS stamp which facilitates the demolding step. Furthermore, the structures are thermally decoupled from the foil since they are made of a different material with a higher temperature stability. In order to obtain good results in the Thermoforming process with this foil assembly, some parameters have to be considered in the UV-NIL process. The thickness of the epoxy layer has to be adjusted to the depth of the surface structures to inhibit a cracking of the epoxy layer during this process, since the cured epoxy material is quite brittle. The maximum tolerable thickness of the epoxy layer depends on the stretch ratio of the foil. For a channel structure with a linear stretch

6. Fabrication of 3D Structures

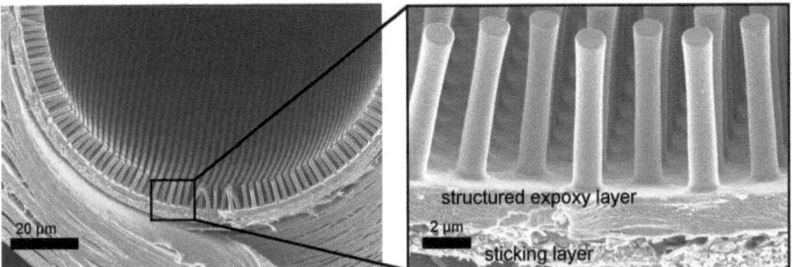

Figure 6.9: SEM image of a cross-sectional view of a thermoformed microchannel. On the right side a SEM image of increased magnification is depicted, showing pillars with a diameter of 1 μm and a height of 7 μm.

ratio of 4, as used in this work, the maximum acceptable residual layer thickness is about 3 μm.

The second point is that the pure epoxy layer does not adhere strongly to the thermoplastic foil and can therefore peel off during the Thermoforming process. Both mentioned problems can be solved by applying the ether component to the epoxy material. The ether component lowers the viscosity as mentioned and leads to a higher adhesion of the epoxy material to the foil. The ether component works as a reactive solvent for the epoxy material and is therefore integrated into the polymer network.

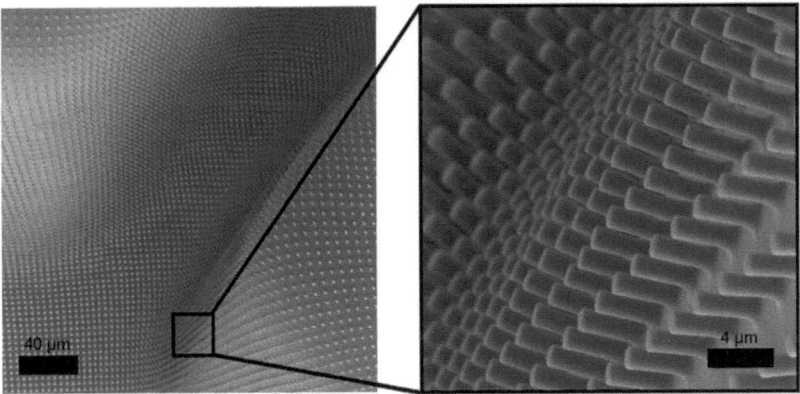

Figure 6.10: The picture shows the entrance of a micro channel (left) and on the right an image of increased magnification is shown. The pillars on the foil have a diameter of 1 μm and a heigth of 4 μm.

In Fig.6.9, a picture showing the foil after the Thermoforming process is depicted. In the magnification, it can be seen that the high aspect ratio structures maintain in good quality after the Thermoforming process. Furthermore, it can be seen that a sticking

67

6.4 Foil Structuring using UV-NIL

layer is formed between the structured epoxy layer and the thermoplastic foil. The formation of the sticking layer can be attributed to the ability of the ether to partly dissolve the polystyrene foil. The foil is partly dissolved on the surface and, after curing the epoxy mixture is bonded to the foil leading to a higher adhesive strength.

In Fig.6.10, a thermoformed foil is illustrated. It can be seen that both, the covering of the microchannel can be done with high accuracy due to the use of higher temperature and pressure and the shape of the surface structures is not affected because of the higher temperature stability of the epoxy material. The main goal of this optimization was to obtain higher aspect ratios on a 3D surface. As mentioned before, the minimum diameter of the structures was 600 nm and the height was 7 μm, therefore the maximum aspect ratio was 11.7. In Fig.6.10 the results of the Thermoforming process for these structures is depicted. On the left side an image of the Si master structure is illustrated. In the middle, a thermoformed foil containing the high aspect ratio structures is shown and on the right side an image of increased magnification can be seen. The nanostructures were replicated in good quality with the described UV-NIL process. Furthermore, the Thermoforming process did not influence the quality of the structures due to the chosen material combination.

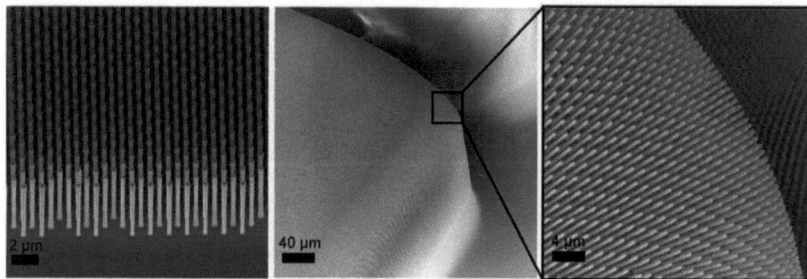

Figure 6.11: Result of the Thermoforming process for nanostructures with a diameter of 600 nm and a height of 7 μm. On the left side an image of the master structure is illustrated. In the middle a thermoformed foil containing the high aspect ratio structures is shown and on the right side an image with increased magnification is shown

In Fig.6.12, SEM images of a cross sectional view of the same structure is shown. The structures could be applied to the vertical side wall of the channel and, even for high aspect ratios, the shape of the nanostructures is not affected over a large area. It was demonstrated, that, due to the optimization of the foil preparation, high aspect ratio structures in the nanometer range could be molded to the sidewalls of micro channels without quality loss. The new foil assembly does not only improve the quality of the micro- and surface structures, but also has influence on the general behaviour of the foil in the Thermoforming process itself. Therefore this multilayer foil behaves differently during the Thermoforming process than a normal thermoplastic foil. In the first part

6. Fabrication of 3D Structures

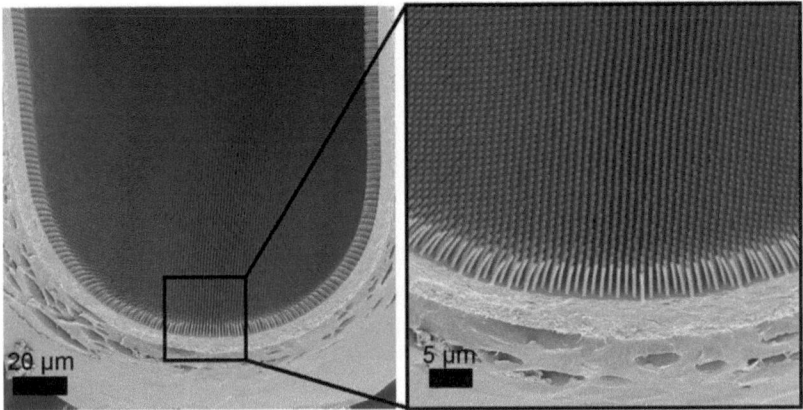

Figure 6.12: Cross sectional view of a thermoformed foil containing high aspect ratio structures. The pillars have a diameter of 600 nm and a height of 7 µm.

of the following section the different behaviour of this foil assembly will be discussed.

In this method of integrating functional surface structures into a micro structure, the integration is done by molding a foil into a microstructure. This foil has a defined thickness which involves some geometrical limitations on the process. These points will be discussed in the second part of the next section.

6.5 Possibilities and Limits

6.5.1 Foil Thickness Distribution

Due to the new foil assembly, the mechanical properties of the foil have changed and, therefore, the behaviour of the foil during the Thermoforming process. In section 2.2 the typical behaviour of a foil during the Thermoforming process was described. The, typically observed thinning of the foil was also noticed in the first test (cp. Fig.6.3). Using this new foil assembly, the thinning of the foil can be reduced and a more homogeneous foil thickness can be observed after the Thermoforming. An image of a cross sectional view of a thermoformed foil, consisting of this multilayer assembly is shown in Fig.6.13. The parameters during the Thermoforming process have been chosen similar to the experiment described in section 6.2.2 (T=105°, p=30 bars). It can be seen that the foil thinning can be reduced drastically.

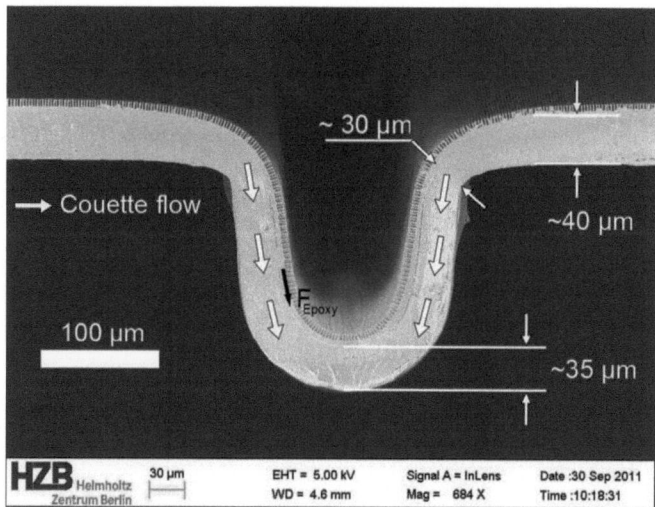

Figure 6.13: Foil thickness distribution of a multilayer foil after thermoforming.

The more homogeneous foil thickness after the Thermoforming process can be attributed to the mechanical properties of the different layers. The thin PDMS membrane on the backside of the foil is highly flexible and has minor influence on the thickness variation, since it adapts to the foil without influencing the thickness. Therefore, the effect of more homogeneous thickness after the Thermoforming process can be attributed to the structured epoxy layer.

6. Fabrication of 3D Structures

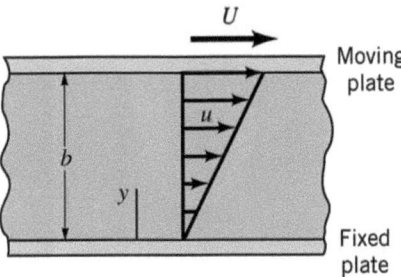

Figure 6.14: Schematic of the couette flow in the foil leading to the homogeneous foil thickness distribution[129].

In order to understand this effect, the physical properties of the epoxy layer play an important role. The epoxy layer forms a cross linked polymer network and has the properties of a thermoset material. As mentioned before (section 2.2.1.4) and as described in [57], thermoset materials need a higher pressure to be formed in a Thermoforming process. This means that the structured epoxy layer has a higher resistance against the pressure than the PS foil. This together with the higher softening temperature of the epoxy material allows the assumption, that the epoxy layer is rigid during the Thermoforming process. Due to this mechanical property of the epoxy layer, the applied pressure during the process leads to a force in the epoxy layer as shown in Fig.6.13. The force that leads to the stretching of the epoxy layer causes a couette flow in the foil leading to the observed foil thickness distribution. This effect is also illustrated in Fig.6.14.

The reduced thinning of the foil also has a positive influence on the molding quality. On one side the deposition of the PDMS membrane on the backside of the foil allows the use of higher temperature and pressure. Together with the effect of reduced thinning of the foil, a higher accuracy of the microstructures can be obtained as the danger of foil cracking attributed to the inhomogeneous thinning effect can be reduced. A foil cracking has not been observed with this foil assembly, even when applying the maximum pressure (60 bar) possible with the Thermoforming tool. The effect of temperature and pressure is illustrated in Fig.6.15.

In the pictures it can be noticed that the pressure has a significant influence on the molding accuracy of a the microstructures. The best results were obtained with a temperature of 115°C and the maximum pressure of 60 bars. Here the micro channel was filled out completely, which means, it has reached the bottom of the channel. A smaller curvature radius cannot feasibly be obtained. It can be seen that even with such high pressures, the thinning of the foil in the channel is quite homogeneous. Therefore, this new approach of foil preparation can be used to obtain a better molding quality in

71

6.5 Possibilities and Limits

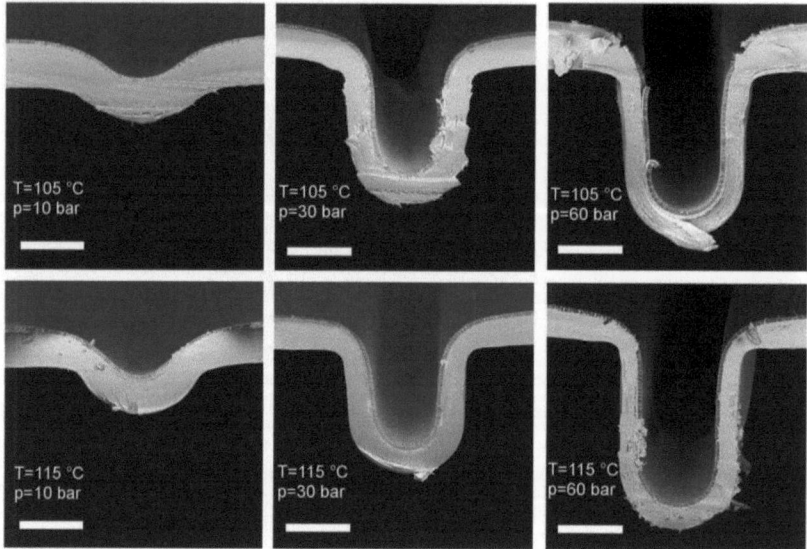

Figure 6.15: Influence of temperature and pressure on the Thermoforming result (scale bar = 100 μm).

the Thermoforming process. For the foil molded with a pressure of 60 bars at 115°C a linear stretch ratio R_L (Eq.2.4) of 3.23 could be obtained (for the calculation the outer foil length was used). This is 81% of the theoretical possible value of 4.

6.5.2 Geometrical Limitations

In the previous section the effects of the new foil assembly on the behaviour of the foil during the Thermoforming process were discussed. It was shown that the new assembly can be used to obtain a higher molding quality of the final structured foil and therefore, the previously mentioned limitations regarding the micro- and the surface structures can be reduced. Other resolution limiting factors on the other hand are attributed to simple geometrical factors and therefore, cannot be solved as easily. These limitations are caused by the foil deformation during the Thermoforming process.

This deformation and the subsequently caused modification of the period of the pillars is schematically shown in Fig.6.16. On the left side, the effect of foil deformation on the pillar structure is illustrated. This effect was investigated among others in [128]. In that work, a hydrophobic channel of PDMS was fabricated by a Hot Embossing process. That process has some limitations, especially regarding the aspect ratio of the nanostructures. These limitations in the manufacturing process could be eliminated

6. Fabrication of 3D Structures

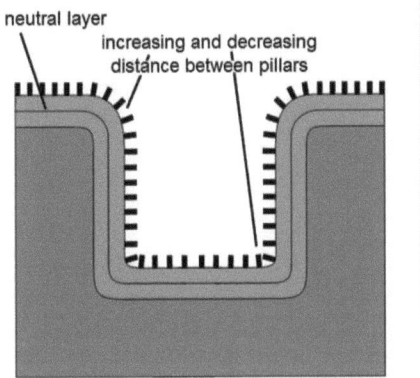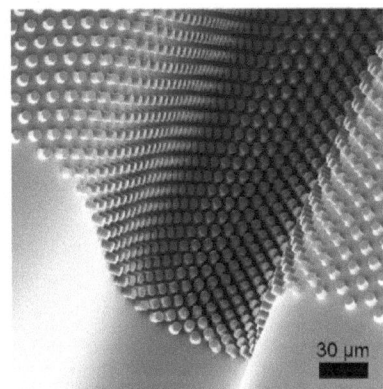

Figure 6.16: During the Thermoforming process a foil deformation takes place as shown on he left side. On the right side a thermoformed foil with pillars with a diameter of 6.4 μm can be seen which visualize the effect of foil deformation.

with the Thermoforming process, using the new foil assembly. The effects which occur because of the foil deformation cannot be overcome with the Thermoforming process. The deformation of the foil leads to different distances of the pillars depending on the area. The distance between two pillars at the bottom border of a channel can be described by following equation [128]:

$$D_1 = \left(1 - \frac{h+t/2}{r+t/2}\right) D_{0p} \tag{6.1}$$

Here, D_1 is the distance between two pillars at a bottom corner after the Thermoforming process, D_{0p} is the original distance of the heads of the pillars before the Thermoforming process, r the curvature radius, t the thickness of the foil including the PDMS membrane and h is the height of the pillars. The curvature radius of the foil at the bottom corner of the channel depends on the used pressure and temperature.

In Fig.6.17 the distance between two pillars after the Thermoforming process is plotted with dependence to the curvature radius for various pillar distances. It can be seen that especially for smaller pillar distances, the values after the Thermoforming process are in the range of a few 100 nm for curvature radii of below 25 μm.

On the right side in Fig.6.16, a SEM image of a thermoformed foil containing pillars with an initial distance (and diameter) of 6.4 μm is depicted. The effect of the modification of the period of the pillars can clearly be noticed. As these structures have a relatively large original distance, they are still separated. Due to optimization in the foil preparation, especially the application of a PDMS membrane, which allows the use of higher pressures and temperatures, a high accuracy can be obtained in the

73

6.5 Possibilities and Limits

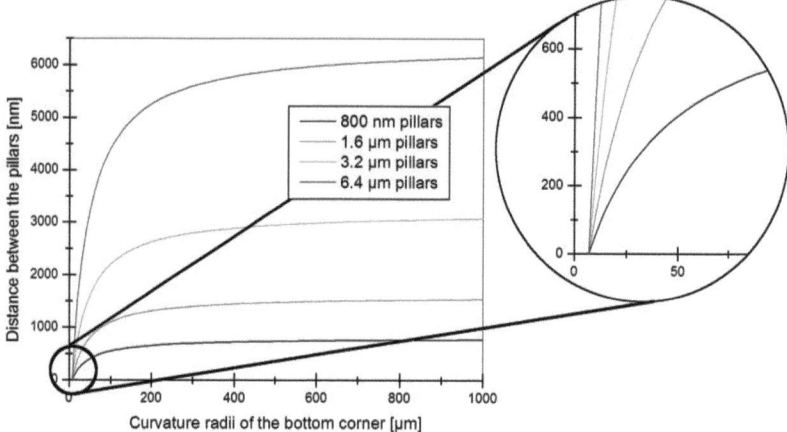

Figure 6.17: Distance between the pillars at the bottom corner of the channel after the Thermoforming process with dependence to the curvature radius for various pillar distances.

Thermoforming process leading to curvature radii of below 25 μm.

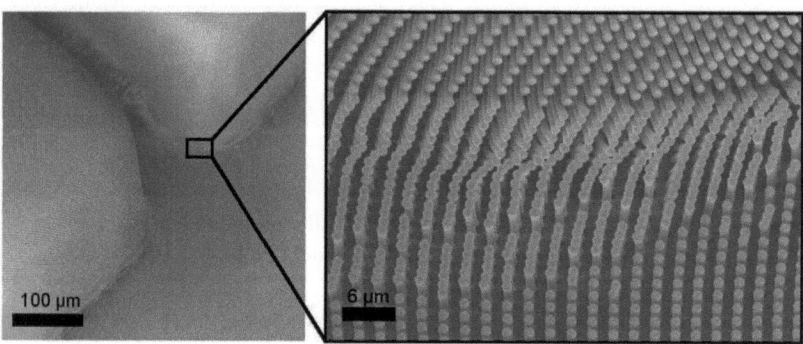

Figure 6.18: SEM pictures of the edge of a thermoformed foil containing structures with an initial distance 3.2 μm.

In Fig.6.18 SEM pictures of a thermoformed foil with a pillar distance of 3.2 μm is depicted. The Thermoforming process was performed with a pressure of 60 bars and a temperature of 115°. The formation of such small radii at the bottom corner of a microstructure leads to a very small distance between the pillars in these areas. During these experiments, a change in the pillar geometry was not observed, therefore it is assumed, that only the areas between the pillars are stretched or compressed. In Fig.6.18(right) the original distance between and the diameter of the pillar heads was 3.2 μm. At the bottom corner of the structure the distance has decreased significantly

6. Fabrication of 3D Structures

until the pillars touch each other. Since this effect is inherent to the method itself, it can not be avoided. But the effect also shows the high accuracy (small curvature radii) that can be obtained in this process.

Other limitations are set by the thickness of the foil. The thickness of the foil determines the minimal structure detail that can be obtained. Foils with a minimum thickness of 20 µm were used in this work. In order to define the minimal structure size that can be obtained with this process, a mold insert containing pillars with a constant diameter of 100 µm were fabricated using micro milling brass. The distance and the height of the pillars was varied in different fields.

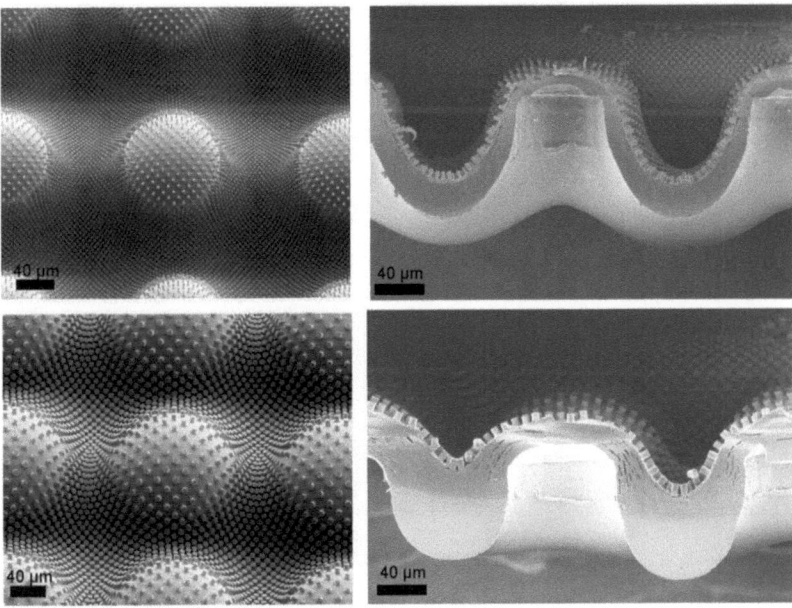

Figure 6.19: Thermoforming result for different foil thickness on a pillar field with a pillar diameter of 100 µm, a height of 200 µm and a distance of 100 µm. Upper row: foil thickness =20 µm, lower row: foil thickness= 50 µm

In Fig.6.19 the result of the Thermoforming process for a pillar distance of 100 µm and a height of 200 µm is depicted. Here a temperature of 105° and a pressure of 60 bars was applied. In the upper row the initial foil thickness was 20 µm. On the left side the effect of foil deformation can clearly be noticed due to the different pillar distances. On top of the resulting hill structure, the pillar distance is significantly larger compared to the valleys in between. In the cross sectional view on the right side, it can be noticed that the foil thickness at the top corner of the pillar is quite thin compared to the rest of the foil. Furthermore, the restoring force in the epoxy layer is bigger in such small

75

6.5 Possibilities and Limits

cavities leading to a higher thickness of the foil in the bottom area. In the lower row of Fig.6.19 the initial foil thickness was 50 μm. Here the polymer flow towards the valleys is even stronger, since the pillar structures touch each other, avoiding further adaptation of the foil surface to the pillar.

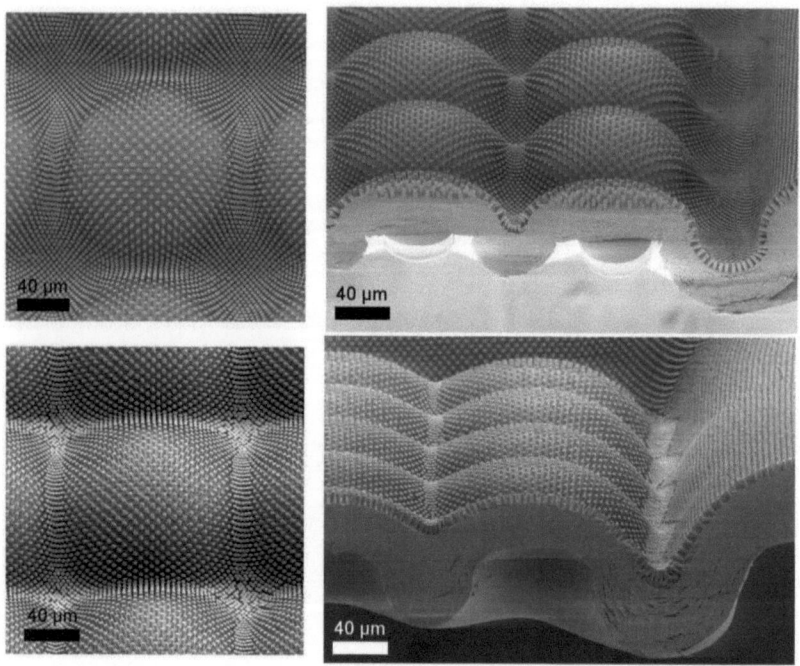

Figure 6.20: Thermoforming result for different foil thickness on a pillar field with a pillar diameter of 100 μm, a height of 100 μm and a distance of 50 μm. Upper row: foil thickness =20 μm, lower row: foil thickness= 50 μm

In Fig.6.20 the results for the same experiment for a pillar field containing pillars with a diameter of 100 μm, a height of 100 μm and a distance of 50 μm are depicted. In the upper row the initial foil thickness was again 20 μm and in the lower row the initial thickness was 50 μm. In all 4 pictures the previous mentioned effects can be noticed in a more extreme way. Especially in the cross sectional view for the 50 μm thick foil in the lower row on the right side, it can be seen that the structured foil surface is only formed a few tens of μm deep into the cavities. At the border of the pillar field the distance is 100 μm. Here the foil can be formed into the cavity a little deeper until the pillars on top of the foil touch each other, avoiding any further foil deformation.

The results from these experiments have shown the limits of this molding technique regarding the structure size. The minimum structure size that can be obtained in good quality depends on the thickness of the foil. For the 20 μm thick foil the pillar structure with a distance of 100 μm and a depth of 200 μm could be formed in an acceptable quality. For thicker foils or smaller pillar distances, no adequate molding accuracy can be obtained. In these cases only a small deformation of the foil surface takes place. Even though the molding of the pillar structure is not good for these cases, the observed effects may be suitable for other applications, like the fabrication of gapless micro lenses.

7 Replication of 3D Structures using Replica Molding

In the previous chapter the Thermoforming process was presented for the fabrication of 3D structured foils. These foils should further be used as master structures in a Replica Molding process for the fabrication of 3D structured polymer parts. It has been shown that Replica Molding processes can be used for the replication of structures like a lotus leaf or optical surface structures [34, 84, 96]. In the first section of this chapter, the development of the Replica Molding process will be highlighted. In this context, the necessary process steps, the minimum resolution, and the possibility of replicating combined micro and nanostructres will be discussed. In the next section, the replication of 3D structures using the thermoformed foils as master structures will be presented. This process chain was used for the fabrication of a 3D structured fluidic system and the functionalization of micro lenses. The developments for the fabrication of these applications are described in the next chapter.

7.1 Replica Molding for Multilevel Micro-Nanostructure Replication

The process of Replica Molding a micro-nanostructured substrate involves four main steps: (1) fabrication of a nano patterned Si substrate; (2) definition of a second level of microstructures on the Si substrate to obtain a multilevel master, (3) obtaining a PDMS stamp from the multilevel master by Soft Lithography and (4) replication of the PDMS stamp in an epoxy based resin by Replica Molding. Each of these process steps is described in detail in the following paragraphs and schematically summarized in Fig. 7.1.

Fabrication of a Silicon Nanostructured Substrate

Arrays of dots, lines and a honeycomb pattern were structured onto a Si substrate by (HE-NIL) using a previously EBL nanostructured silicon stamp. These processes were performed as described in Chapter 3 and 4. For the imprint a 80 nm thick PMMA layer was spincoated onto the substrate and the nanostructured silicon stamp was imprinted into the PMMA layer at 190°C with a pressure of 10 MPa for 5 min. After demolding, the residual layer was removed and the PMMA nanopatterns were

7.1 Replica Molding for Multilevel Micro- Nanostructure Replication

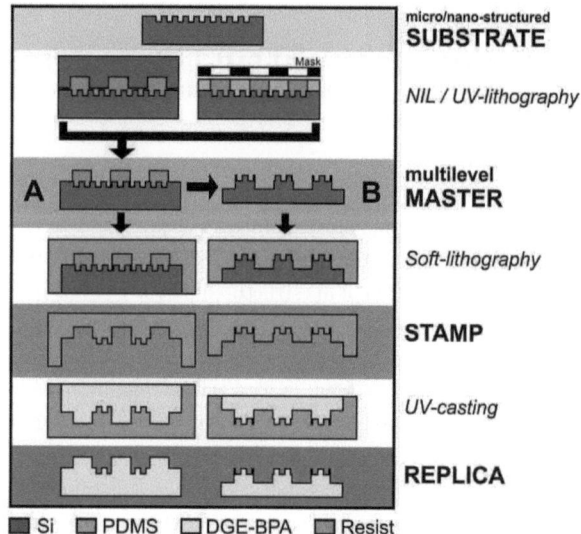

Figure 7.1: Schematic diagram of the Replica Molding process used to replicate the multilevel masters (A with nanostructures at the bottom or B with nanostructures on top).

transferred to the silicon substrate by pattern transfer via Lift Off and RIE (chapter 4).

Fabrication of the Multilevel Micro-Nanostructured Master

In order to obtain a master combining the former nanostructures with a second level micro pattern, an additional HE-NIL step with a stamp consisting of lines and spaces was performed on the nanostructured silicon substrate. The lines and spaces have widths of 800 nm and 1.6 μm. A 1.4 μm thick layer of PMMA was spincoated over the substrate and the microstructured silicon stamp was imprinted at 190°C with a pressure of 7.5 MPa for 5 min. After demolding, the residual layer was removed by anisotropic etching in oxygen plasma. At this point two options arise: the combination of the PMMA and silicon substrate can be used directly as a multilevel master for the PDMS molding or alternatively, PMMA can be used as an etch-mask for the silicon substrate. In the first case, nanostructures remain at the bottom of the silicon substrate whereas in the second case they will be placed on top of the microstructures once the PMMA is removed. Besides the possibility of using two HE-NIL steps, the structuring of the microstructures can also be performed by UV Lithography.

7. Replication of 3D Structures using Replica Molding

Obtaining of PDMS Stamp

For the fabrication of the PDMS stamp, no additional treatment of the multilevel master structure was necessary. The multilevel master was placed inside a metal frame. The PDMS prepolymer was mixed according to the manufacturer specification [86], degassed for 5 min to avoid air bubbles and poured over the master. The frame was covered by a metal plate which was put in contact with the PDMS mixture to define a flat surface on the back of the stamp and the assembly was pressed with a force of 3 kN and heated to 70°C for 30 min for the curing process. After curing, the PDMS stamp was peeled off from the master.

Replica Molding with an UV-Curable Epoxy-Based Resin

The replication of the structures was made with a mixture of diglycidyl ether of bisphenol A (DGE-BPA) and 3 wt-% of triaryl sulfonium salt (TAS) as a photoinitiator (Aldrich) [48, 51, 52]. To achieve a lower viscosity, and therefore ease the cavity filling, the mixture was heated up to 70°C [130]. This mixture was then cast onto the PDMS stamp. A thin foil of polycarbonate was placed on the stamp to ensure a plain back surface after the replication. The exposure was performed with a dose of 1200 $\frac{mJ}{cm^2}$. The cured structured part was easily peeled off the PDMS stamp and released from the polycarbonate foil.

Three different designs were patterned into the silicon substrate by HE-NIL, lines (50 nm line width), dots (75 nm diameter) and a honeycomb pattern (110 nm web width). The structures were etched 80 nm deep into the substrate. As a preliminary step, the Replica Molding process capabilities were tested by using the one level nanostructured substrate as a master. These structures are suitable to qualify the accuracy and the sharpness of the replication as they consist of very small features, some of them with corners (honeycomb pattern). A PDMS stamp with a thickness of 5 mm was reproduced from this substrate and used to replicate the structures in DGE-BPA. Even though the features have small dimension, no sticking problems occurred during demolding due to the hydrophobic character and the elasticity of the PDMS. The accuracy of the replications was verified by SEM inspection of the Si master and the resin replicas. SEM images of both, the silicon master structure and the resin replicas for two of the three fabricated designs are shown in Fig. 7.2. The trenches and the web of the honeycomb pattern in the resin have a width of 49 nm and 112 nm respectively. The dimensions of both samples were measured using the SMART SEM program from Zeiss and the comparison shows a difference of less than 5 % in the XY plane.

7.1 Replica Molding for Multilevel Micro- Nanostructure Replication

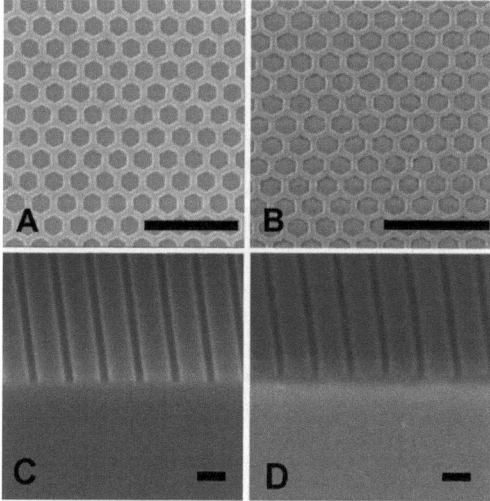

Figure 7.2: SEM images of the master structure of a honeycomb pattern (A) with a web width of 110 nm and B) the replication in DGE-BPA (scale bar = 2 μm). Images C) and D) illustrate a cross sectional view of the 50 nm wide trenches master structure and its DGE-BPA replication, respectively (scale bar = 400 nm).

Multilevel Replication

The replication of the combined micro and nanostructures was done with the two alternatives of multilevel master (A and B in Fig. 7.1). Therefore, a pattern of lines and spaces with a 800 nm and 1.6 μm pitch was defined by NIL in a 1.4 μm thick PMMA layer. For case A, the PMMA thickness defined the depth of the second microstructure layer. For case B, the PMMA layer was used as a mask to etch the structures 1 μm deep into the silicon substrate by RIE. Thus, case B offers the possibility of fabricating a master with deeper microstructures, as the RIE process is not limited to the thickness of the PMMA layer. The master structures were first replicated in PDMS to obtain the stamp and then replications in DGE-BPA were made. Fig.7.3 shows SEM images of the nanostructured substrate (7.3A-C), the multilevel master structure fabricated according to method B (7.3D-F) and replicas of the master structure, processed using multilevel master B (Fig.7.3 G-I).

The master fabrication methods differ in the limitations of the achievable depth and structure quality. In case A, the height of the lines is limited by the thickness of the PMMA. The maximum height achievable by this method is lower than the thickness of the PMMA layer, as the residual layer must be removed in order to obtain the nanostructured bottom. In case B the height of the structures is just limited by the selectivity of the PMMA to the silicon in the etching process. To achieve higher struc-

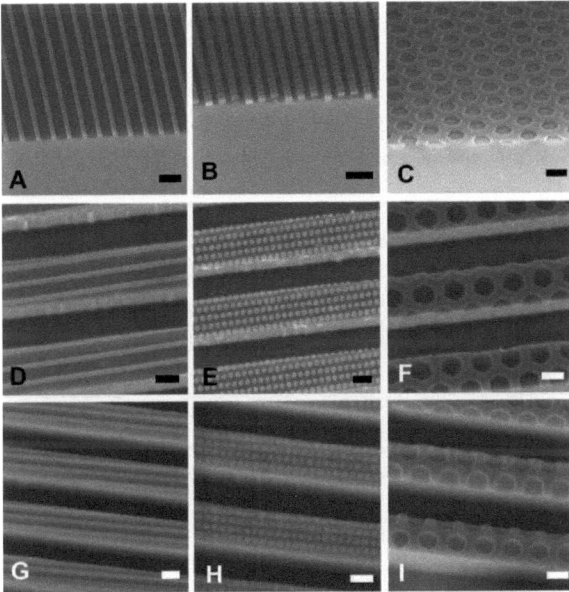

Figure 7.3: SEM images of the nanostructured substrate A)-C) fabricated by NIL. Images D)-F) show multilevel master structures, fabricated by method B and G)-I) show SEM images of the corresponding replications in DGE-BPA (scale bar in images = 400 nm).

tures, a Lift Off step can be used to define a metal etch mask which offers a higher selectivity in the etch process and therefore allows higher multilevel structures to be produced.

7.2 3D Structuring of Polymer Parts

In section 6.2 the Thermoforming tool was presented. In the Thermoforming tool, a ring was integrated for the fabrication of a PDMS stamp. The ring was used to cast the liquid PDMS prepolymer into and it defines the final thickness of the PDMS stamp. The process for fabricating a 3D structured polymer part out of a structured foil is illustrated in Fig.7.4.

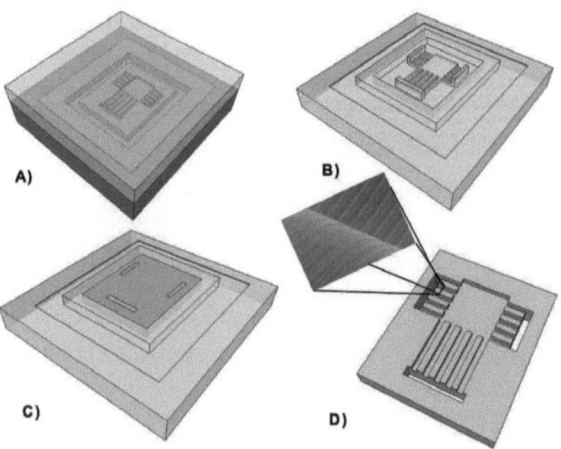

Figure 7.4: Replica molding process for 3D polymer replications. A PDMS stamp is fabricated by casting the prepolymer (light blue) over the structured foil A). The PDMS stamp is demolded B) and the resin (yellow) is cast over the PDMS stamp C). D) The resin is cured by UV exposure and demolded from the stamp.

In the first step after the Thermoforming process, the upper part of the tool is detached in order to get access to the structured foil, then the PDMS prepolymer is cast into the the ring, covering the thermoplastic foil (Fig.7.4A)). The tool is put into an exsiccator and vacuum is applied to achieve a complete filling of the structure. The filled tool is covered by a plate to define the backside surface of the stamp. The curing of the PDMS is done as described before (Chapter 3). After curing, the PDMS stamp can be peeled off (Fig.7.4B)) without damaging the structured foil. The replication of polymer parts is performed as described earlier. In order to fabricate polymer parts, the integration of a frame in the PDMS stamp can be helpful for the casting process in order to remove excessive material and facilitate the demolding process. This frame must be integrated into the mold insert (cf. Fig.6.2). In Fig.7.5, the result after each step of the fabrication process is illustrated. Here SEM pictures of a cross sectional view of a structured foil (left), the corresponding PDMS stamp (middle) and an epoxy replica (right) are depicted. The channel in this picture has a height of 300 μm and

7. Replication of 3D Structures using Replica Molding

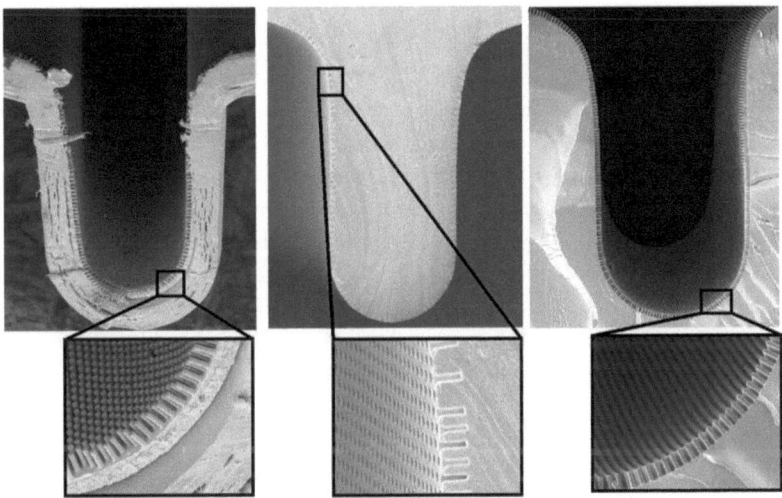

Figure 7.5: SEM images of the process steps beginning from the structured foil (left), the fabricated PDMS stamp (middle) and the resulting polymer replication (right).

a width of about 100 μm. The pillars have a diameter of 1 μm and a height of 7 μm. A SEM picture of a replicated structure is depicted in Fig.7.6, the pillars in this replication have the same dimensions as in Fig.7.5.

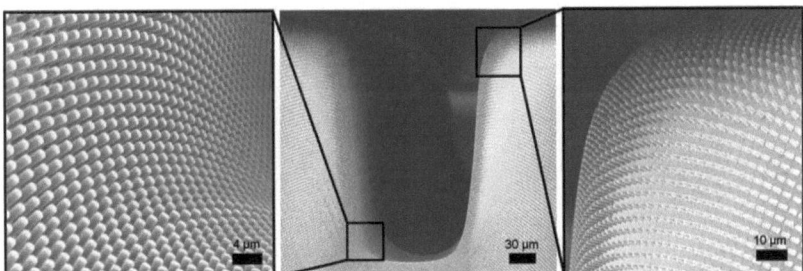

Figure 7.6: SEM pictures of a replicated structure. The channel in the middle has a height of 300 μm and a width of 100 μm. Details of increased magnifications are illustrated on the right and on the left side of the images.

In the replication, it can be seen that the structures were replicated with high fidelity. Due to the flexibility of the PDMS, the structure could be demolded even from sidewalls and edges without damaging them. The possibility of replicating undercuts in polymer devices is an unique feature of Soft Lithography processes [16] which is also the main difference when compared to Hot Embossing or Injection Molding where hard mold inserts are used and, due to the uniaxial demolding step, undercuts cannot be

7.2 3D Structuring of Polymer Parts

fabricated.

Since the epoxy material is quite brittle (tensile strength of epoxy material is typically in the range of 55 MPa), a single pillar element can only be bend a few μm before tearing off. This can be calculated from the Euler-Bernoulli beam theory. Therefore, the non-destructive demolding of such a polymer part can mainly be attributed to the elasticity of the PDMS material, which leads to stretching of the stamp during the demolding process. Assuming that the volume of the PDMS maintains constant (Poisson ratio 0.5), the thickness of the a single stamp protrusion decreases during demolding as the stretching of the stamp leads to an increase of length. This effect occurs sequentially over short distances (about 10 μm). Therefore, the PDMS stamp detaches from the sidewalls of a channel or cavity working its way from the top edge down to the bottom.

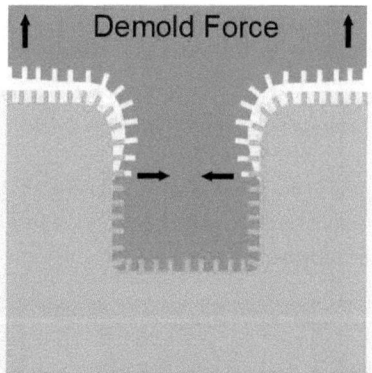

Figure 7.7: Schematic image of the demolding mechanism during Replica Molding.

This effect leads to non-destructive demolding in the Replica molding process. Due to its thinning, the PDMS stamp moves perpendicular to the sidewalls, thus helping to demold the structures on a vertical surface as illustrated in Fig.7.7.

Fig.7.8 shows SEM pictures of a cross sectional view of a replicated structure containing pillars with a diameter of 600 nm and a height of 7 μm. The images illustrate that even such high aspect ratio structures can be demolded.

In this section, the combination of Thermoforming and Replica Molding was used to fabricate 3D structured polymer parts. Due to this optimized foil structuring technique using UV-NIL and the PDMS membrane on the backside of the foil, a high structure fidelity could be achieved in the Thermoforming process. Furthermore, high aspect ratio nanostructures could be formed on the sidewalls of microchannels on a thermoplastic foil. The presented Replica Molding process could be used to transfer these structures

7. Replication of 3D Structures using Replica Molding

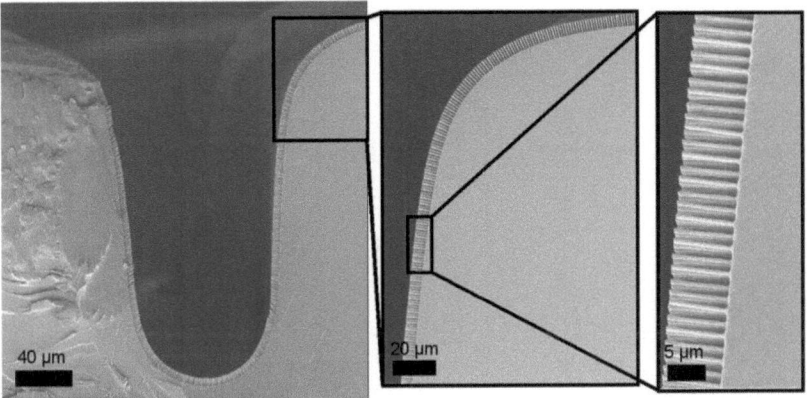

Figure 7.8: SEM picture of a cross sectional view of a replication containing pillars on the channel side walls with a diameter of 600 nm and a height of 7 μm.

into a polymeric device. This process chain was used for the functionalization of microlenses; more precisely, moth eye structures working as a anti-reflection layer were applied to a microlens array. Furthermore, a fuel delivery system was fabricated using the presented process chain. The developments for these applications are presented in the following chapter.

8 Applications for 3D Structured Polymerparts

8.1 Anti-Reflection Structures for Microlenses

Microlenses are an important part of many engineering devices such as optical sensors [131], LEDs [132] or camera modules [133–135]. Especially the field of low cost camera modules requires technological solutions to face the increasing consumer demand for enhanced feature sets and functionality in mobile devices. More than a billion cell phones were made and shipped per annum in recent years [135]. Therefore, wafer scale manufacturing has attracted the attention of suppliers of such devices. One of the most promising approaches to fulfilling the needs of this growing demand are imprint processes [132–134, 136, 137]. In Fig.8.1, an overview of a typical fabrication process using an imprint lithography method is depicted [138].

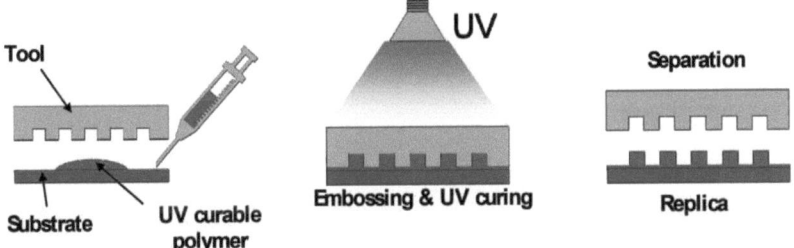

Figure 8.1: Fabrication of optical modules using an imprint lithography process (REEMO®),(source:[138])

In this process, a stamp made of PDMS is produced in order to replicate the lenses in an UV based replication process. Various methods can be used to obtain the master structure for the fabrication of the PDMS stamp. The most widely used processes for the master fabrication are reflow processes [136, 139–141], contactless embossing of microlenses [137] or diamond turning [142]. In order to improve the optical function of the lenses an anti-reflection layer is deposited on the lenses. One possibility of obtaining anti-reflection properties is the deposition of a multilayer coating using vacuum evaporation processes [143]. The deposition of such a multilayer coating requires further processing steps which makes the blooming of lenses a cost–intensive procedure. In such a multilayer assembly, the refractive indices and the thickness of the layers are chosen in a way to adjust the refractive index of the substrate to air, resulting in the

89

8.1 Anti-Reflection Structures for Microlenses

anti-reflective properties of the coating. Using such coatings very low values (below 0.5 %) for the reflection can be obtained [144]. However, the wavelength dependence of the refractive index makes it difficult to obtain a broad band anti-reflective effect [144]. Another option is using nanostructuring techniques. It has been demonstrated that moth eye structures can be used to improve the performance of solar cells or displays [145–147]. The surface relief of the moth eye structure results in a gradual change of the refractive index from one medium to another. This adjustment of the effective refraction indices results in the anti-reflective properties of such a surface [146]. A periodic surface relief on subwavelength scale has shown good anti-reflective properties [146]. An advantage of such a structure compared to multilayer coatings is that they work for a wide range of wavelengths, allowing for the fabrication of broad-band anti-reflective coatings [144]. The optimal design of the moth eye structure can be calculated from the effective height, the spacing, the shape and the wavelength of the light [144]. When optimizing the structure design, values for the reflectance of 0.2 % can be obtained [11] and, therefore, such structures have similar anti-reflective properties when compared to the multilayer assembly. However the fabrication of moth eye structures on a 3D shaped substrate, using standard MEMS processes, is quite complex which makes the integration of such structures into microlenses difficult. The processes described in the previous chapters were applied to the integration of moth eye structures into a PDMS stamp for the replication of functionalized microlenses using a UV-NIL process. The necessary process steps are described in detail in the following paragraphs and are summarized in Fig.8.2

8.1.1 Foil Structuring

The master structure used in this work was obtained from the Fraunhofer Institute for Solar Energy Systems, Freiburg. The structures were fabricated using Interference Lithography followed by an electroplating step [148]. This master structure was used to fabricate a PDMS stamp for UV-NIL as described in section 3.2. The foil structuring was performed with an epoxy mixture containing bis - cycloaliphatic epoxide (BCE), divinylether of triethylene glycol (DVE-TEG) and triaryl sulfonium salt (TAS) as initiator. In this experiment, the structures have a height of about 200 nm. The thickness of the epoxy layer had to be adjusted in order to inhibit a cracking of the cured layer in the following Thermoforming process. To obtain a sufficiently thin epoxy layer the proportion of DVE-TEG in the mixture was increased to 30 wt-% and the velocity during spincoating was set to 5000 rpm. The UV-NIL step was performed as described before. This structured foil was further used as a substrate in the Thermoforming process to cover a lens array.

8. Applications for 3D Structured Polymerparts

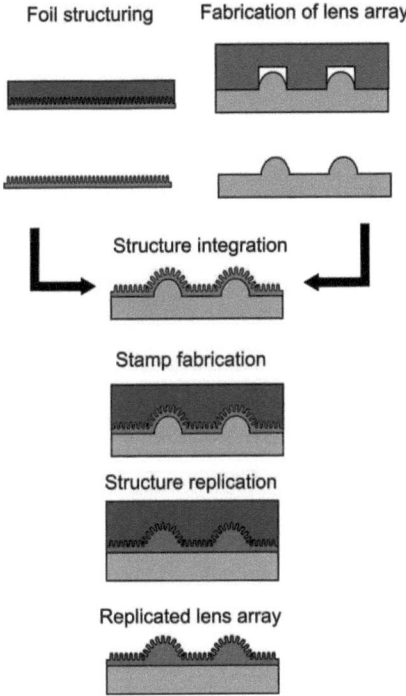

Figure 8.2: Schematic diagram for the integration of the moth eye structures into the PDMS stamp and the replication of the functionalized lens array.

8.1.2 Fabrication of the lens array and replication of functionalized lenses

The fabrication of a lens array was performed using contactless embossing of microlenses (CEM), a method presented in [137]. Therefore, an imprint master had to be fabricated in the first step. The master contains cylinders with different diameters and heights and was fabricated by micromilling of aluminium. The diameter was varied from 1 mm to 0,4 mm in steps of 0,1 mm.

The depth of the cylinders was set to double the size of the diameter. The f-number of the lenses can be adjusted by choosing the parameters during the imprint as described in [137]. The fabrication of the lens array which was used as a master for the Thermoforming process was performed using a PMMA substrate and the imprint was done with a temperature of 160°C, a force of 6 kN (2.4 MPa) and a holding time of 5 min. These parameters result in the pillar height being larger than the lens sag. Since the lens array was used as a master structure for the Thermoforming process,

91

8.1 Anti-Reflection Structures for Microlenses

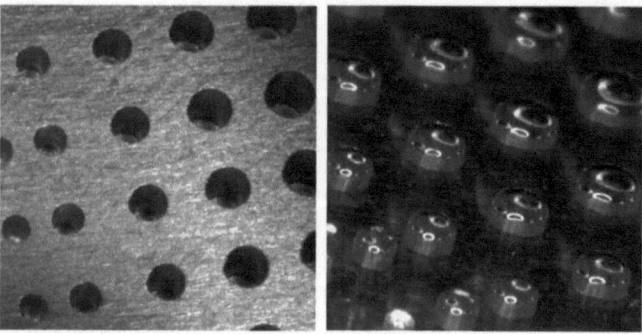

Figure 8.3: Master structure for the fabrication of a lens array (left) and replicated lens array (right).

the additional height of the pillars was compensated for by the thickness of the foil. In Fig.8.3, a digital microscope image of the micromilled imprint stamp and the imprinted lens array are shown. After the imprint the lens array was cut to the size of the inlay in the Thermoforming tool in order to use the PMMA master as a mold insert for the Thermoforming process. The Thermoforming process was performed using a PS foil. The glass transition temperature of the PMMA is 105°C, therefore, the PMMA mold insert could be used directly in the Thermoforming process without additional treatment. As the temperature of the Thermoforming process was set to 107°C with a pressure of 15 bar, no damage was caused to the PMMA master and the master could be used several times. The foil containing the functionalized lenses was further used as a master structure for the fabrication of a PDMS stamp as described in the previous chapter.

8.1.3 Replication Results and Conclusion

The function of the moth eye structure was characterized by UV/VIS spectroscopy of a 1 mm thick epoxy replication containing structured and unstructured areas.

A UV/VIS spectra was taken from both areas using the Lambda 950 equipment from Perkin Elmar. The results of the measurement are illustrated in Fig.8.4. A higher transmission can be achieved over the entire visible range on the structured area compared to the unstructured area. At wavelengths below 420 nm, two effects can be observed. First, it can be seen that the difference between structured and unstructured area gets smaller; this is caused by the dimensions of the structure (i.e. the period and the height of the structure) and second, it can be seen that absorption takes place at these wavelengths leading to a decrease of the transmission. This effect

8. Applications for 3D Structured Polymerparts

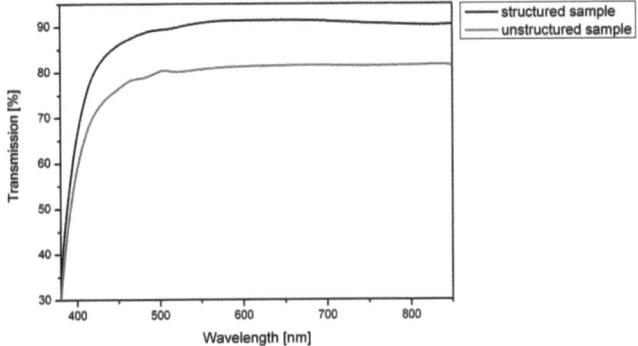

Figure 8.4: Results from the UV/VIS spectroscopy for a structured and a unstructured replication.

is caused by the epoxy material. The initiator of the material is sensitive in the UV range and therefore the absorption in this area increases. However, in the visible range (400 nm - 850 nm) an increment of 10 % in transmission could be achieved with the moth eye structures for this material. The transmission of a material depends on the refractive index, the incidence angle and the absorption of the material. This result demonstrates the qualitative improvement that can be obtained by applying moth eye structured to a surface. In this respect, it is obvious that an improvement of micro lenses can be obtained with moth eye structures. In Fig.8.5, a digital microscope image

Figure 8.5: Structured foil, PDMS stamp and replication in epoxy containing the moth eye structures

of the foil containing the micro lens array with the moth eye structures, the fabricated PDMS stamp, and a replication in epoxy material are depicted. In all images, a blue shimmer, which is caused by the moth eye structures on the lenses, can be noticed. In the UV/VIS spectrum (Fig.8.4) of the replications, it is shown that the anti-reflection

93

8.1 Anti-Reflection Structures for Microlenses

effect of the moth eye structure is working for wavelengths above 420 nm.

Figure 8.6: Digital microscope and SEM image of structured foil containing the moth eye structures.

The reflection properties of such a structure can be influenced by various factors like the period, grating design and aspect ratio [148]. In this case the higher reflectance of wavelengths smaller 420 nm results in a blue shimmer of the foil. The fact that the foils show the same blue shimmer as the original structure reveals that the moth eye structures were replicated with high accuracy. SEM pictures of the results from the Thermoforming process are depicted in Fig.8.6. The moth eye structures on the lens

Figure 8.7: Digital microscope and SEM image of the replicated lens array containing the moth eye structures.

array maintain their shape because of the higher temperature stability of the epoxy material. A PDMS stamp was fabricated using the structured foil as a master structure. Since the lens array should be replicated in a thin epoxy material, the PDMS stamp was not used as master for the Replica Molding process but as stamp for an UV NIL step. The epoxy mixture that was used for this replication contains only the epoxy

8. Applications for 3D Structured Polymerparts

component (BCE) to obtain a higher viscosity and a corresponding thicker layer on the substrate after the spincoating. The epoxy was spincoated with a velocity of 800 rpm onto a polycarbonate substrate. The fabricated PDMS stamp was then brought into contact with the epoxy layer on the substrate. The liquid epoxy layer fills the cavities of the PDMS stamp and then the assembly was heated up to lower the viscosity of the epoxy which resulted in a complete filling of the moth eye structures as well. Due to applying temperature before exposure, the remaining residual layer could also be minimized. The exposure of the epoxy mixture was performed as described with a dose of 900 $\frac{mJ}{cm^2}$. The results of the UV-NIL process can be seen in Fig.8.7.

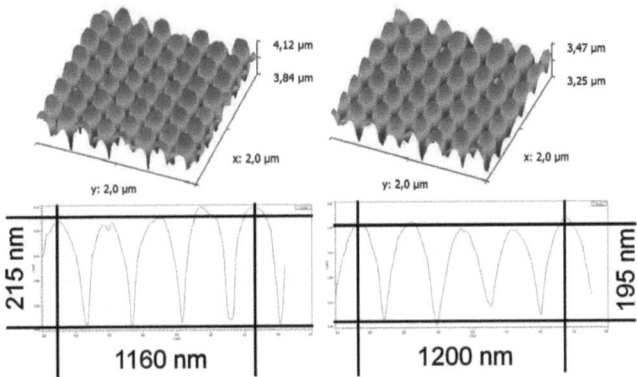

Figure 8.8: Results of the AFM measurement of the original master structure (left side) and the replicated lens array (right side).

The transfer fidelity of the process was verified by AFM measurement. Therefore, images of the original master and the final replication were taken and compared. The AFM image of the replication was taken from the flat, structured surface at the border of the lens array. The results of the measurement are depicted in Fig.8.8. On the left side the results from the measurement of the original master structure is depicted. The measured height of the master structures of about 215 nm and a period of 290 nm is in good accordance with the data provided by the manufacturer (200 nm in height and a period of 280 nm). After the entire replication process a structure height of about 195 nm and a period of 300 nm was found. The smaller height of the structures and the greater period was measured at various positions of the replication. These changes can be caused by various factors. The difference in height can be caused by shrinkage of the epoxy material during the foil structuring and the replication process. The larger period can be caused by the foil stretching during the Thermoforming process. Since the changes in period and height are very small, the optical properties are hardly affected by them.

8.2 Fuel Delivery System for Micro Fuel Cell Applications with Passive Feed

The demand for miniaturized power sources has increased rapidly in the past few years due to the spread of portable electronic applications. Micro fuel cells have gained interest as a portable power source due to their high theoretical efficiency and power density [149]. Fuel cells promise to provide more reliable and longer portable power than batteries as their energy is stored as a fuel instead of being a part of the power source [150–152]. The performance of a fuel cell is described mainly by its Intensity-Voltage (I-V) curve. An ideal fuel cell would supply any amount of current at a constant voltage determined by thermodynamics [153]. In practice, there are some limitations regarding the voltage that can be obtained. There are three main effects causing the limitation in the performance of a micro fuel cell. First, there are activation losses due to the electrochemical reaction; secondly, the ohmic losses caused by the ionic and electronic conduction; and, lastly, there are concentration losses due to mass transport [153]. Therefore, the voltage that can be obtained from a single cell is usually to small to power an electronic device. In order to provide a higher voltage, fuel cells are typically connected in series resulting in a stack. For active fuel cells, a stack can be built by using bipolar plates as current collectors and for the interconnection of various fuel cells and to deliver the reactants to the reaction site. In active fuel cells, the fuel delivery and conditioning is handled by external pumps and mixers [150, 154]. Therefore this strategy can be applied very efficiently to these fuel cells, as the flow rates of reactants can be adjusted to control their concentration in the system and the produced by-products are removed continuously [153]. However, the need of these external systems makes active fuel cells more complex and reduces their availability for portable electronic devices. In passive fuel cells, no external accessories are needed. Passive fuel cells use a reservoir filled with a fixed volume of fuel that is consumed until its empty; therefore, fuel distribution relies on diffusion. Passive fuel cells not only offer the advantage of a simple and compact systems but also makes it possible to eliminate parasitic power losses which occur in active systems. In passive systems the accumulation of by-products has a negative effect on the performance of the fuel cell, e.g. when the concentration of the carbon dioxide generated in a direct methanol fuel cell exceeds a certain level, the performance is deteriorated because of the reaction sites blockage and mass transport difficulties [155, 156]. Therefore, a stack configuration of passive fuel cells requires a management of the involved chemical species in order to guarantee the correct device operation.

In this work, a fuel delivery system was fabricated by applying hydrophobic and hydrophilic areas to a microfluidic system. The functionalization of the different areas was obtained by surface structuring. The fabrication of this fluidic system was developed in collaboration with the micro fuel cell group at the Centro Nacional de Microelectronica (CNM) in Barcelona. To fabricate the fuel delivery system various attempts were made; these will be described in detail in the following section.

8.2.1 Hybrid Fuel Cell and System Integration

The fuel delivery system was designed for a fuel cell presented in [153]. The fuel cell consists of a membrane electrode assembly (MEA)sandwiched between two micromachined current collectors which are integrated into an acrylic casing. The whole assembly is illustrated in Fig.8.9. The current collectors have a width of 10 mm and a length of 14 mm. These dimensions define the size of the fuel delivery system.

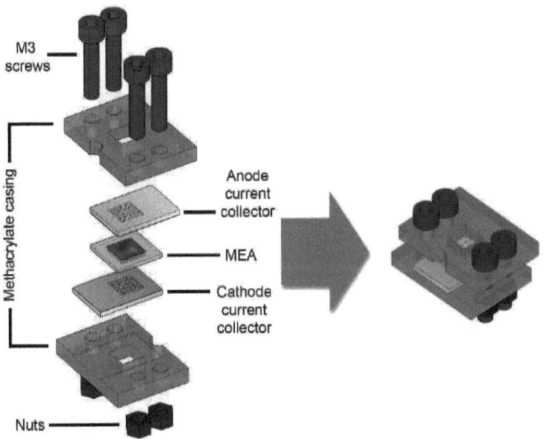

Figure 8.9: Hybrid fuel cell parts and assembly [153].

In this fuel cell, a solution of methanol (MeOH) in water is used as hydrogen supplier. A typical solution for a passive methanol fuel cell is 2 molar (2M)[153]. The hydrogen molecule is broken by the catalyst at the anode side of the fuel cell into protons and electrons. Electrons are collected at the anode side, establishing a current that can be derived to an external load in order to generate the required power. Meanwhile, protons cross the electrolyte membrane towards the cathode where they recombine with oxygen molecules from air and the electrons coming from the external circuit to produce water. The overall reaction of the direct methanol fuel cell is:

$$CH_3OH + 1\frac{1}{2}O_2 \longrightarrow 2\,H_2O + CO_2$$

8.2 Fuel Delivery System for Micro Fuel Cell Applications with Passive Feed

Up to now, the filling of the fuel cell is performed using a pipette as illustrated in Fig.8.10. In this way the characterization of the micro fuel cell could be done [153], but for further system integration and in order to use the fuel cell in an application a fuel delivery system is highly desirable.

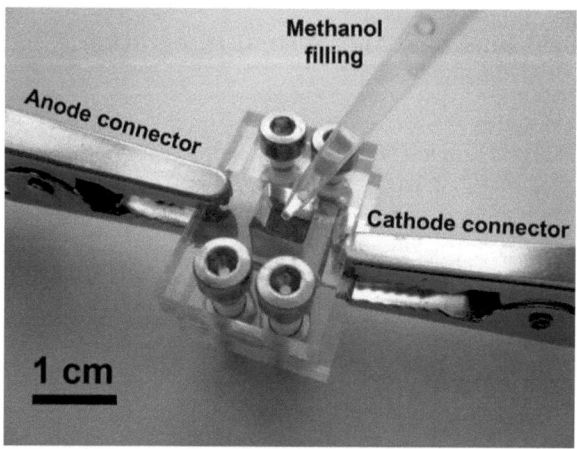

Figure 8.10: Filling of the anode reservoir of the micro fuel cell [153].

The fuel delivery system should distribute the fuel to the cell and allow for the exhaustion of the produced gas by-products. These two functions should be obtained by adjusting the wetting properties in specific areas to the specific needs. The fuel distribution should occur through hydrophilic channels, while the produced CO_2 should be exhausted through hydrophobic channels. The wetting properties of the different areas are defined by selective patterning. The basic design, the integration and the function of the fuel delivery system is schematically illustrated in Fig.8.11.

Figure 8.11: Basic design of the fuel delivery system (left), integration into the fuel cell (middle) and function (right) [153].

8.2.2 First Version of the Fuel Delivery System

In the first version of the fuel delivery system a multilevel structure was fabricated using a lines and spaces structure for the functionalization of the channels. It is known

8. Applications for 3D Structured Polymerparts

that an ordered pattern of lines can also define the contact angle of a liquid [42]. In this particular case, the contact angle of a fluid is different in the parallel and the orthogonal directions in reference to the lines. Furthermore, it could be shown that the difference in the contact angle between parallel and orthogonal direction is larger for narrow linewidths [42]. This effect is interesting for the fuel delivery system application because a difference in the hydrophobicity of the channels could be obtained by aligning the hydrophilic channels to nanostructured lines in the substrate. Consequently, the other channels will be aligned in a direction orthogonal to the lines, becoming hydrophobic. In order to determine the most convenient structure size for the application, resin substrates with nanostructures of different dimensions were characterized by contact angle measurements. The measured structures consisted of lines and spaces with a width and pitch of the lines of 800 nm, 1.6 μm, 3.2 μm and 6.4 μm. These structures were etched 1 μm deep into Si and replicated in DGE-BPA following the process described in the previous sections. The contact angle measurements were performed

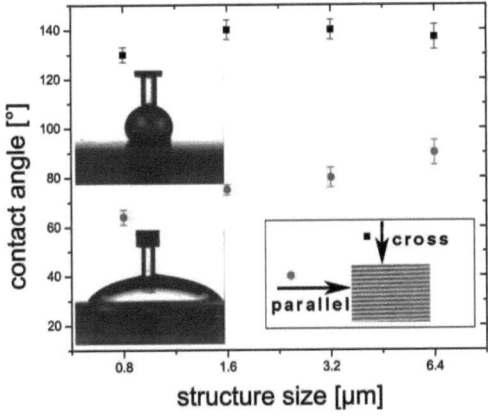

Figure 8.12: Contact angle measurement values obtained from the line-structured surface of replicated resin using a 2 molar solution of methanol. The angle magnitude depends on the dimensions and direction of the structures

with an EASY-Drop-Standard equipment from Kruess GmbH. The measurements were realized with a 2 molar solution of methanol, a typical concentration used for this kind of fuel cells. As similar behaviour was observed for pure water, in the following a high contact angle obtained with the methanol solution is referred to as hydrophobic. A detailed description of the procedure followed to perform the contact angle measurement can be found in [157, 158]. The results of the contact angle measurements are summarized in Fig. 8.12, where the mean values of the measured equilibrium angles are plotted. The graph shows that the measured values of the contact angle of a drop in the direction parallel to the lines range from 65° to 85° for the different pitch/width lines. For the orthogonal direction, the contact angle values ranges from 130° to 140°. As

99

8.2 Fuel Delivery System for Micro Fuel Cell Applications with Passive Feed

the application requires the maximum difference between hydrophobic and hydrophilic behavior, the structures of 800 nm and 1.6 μm wide lines were chosen, as the difference obtained with these structures was more than 70°. Besides these measurements, samples of plain resin and unstructured SU-8 were also measured (not shown). These last measurements are important because the device should be bonded to SU-8 to achieve closed channels. The measured contact angle formed by the methanol solution on both materials was about 65°, corresponding to a hydrophilic behavior.

8.2.2.1 Fabrication Process

The fabrication of the master started with the UV Lithography of the 800 nm lines, followed by a Ni deposition and a Lift Off process. Then, using RIE, the silicon surface was etched to a depth of 1 μm. The remaining Ni layer was finally removed by HCl dip. After cleaning and drying, the wafer was exposed to oxygen plasma (50 sccm, 30 W for 5 min) to remove any organic residues and then dehydrated at 200°C for 10 min. A lithography of a 100 μm thick SU-8 layer was performed to define the micro channels, the entrance hole, and the active area. For this step it was important to align the mask

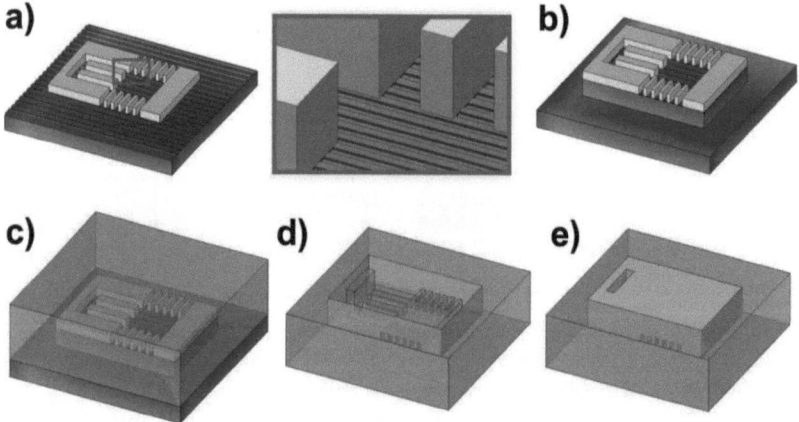

Figure 8.13: Fabrication process of the fuel delivery system. a) SU-8 lithography on nanostructured substrate and the magnified view of the indicated area; b) Silicon laser cutting and bonding on plain substrate; c) Soft-lithography of multi level master; d) PDMS stamp; e) UV Casting.

so that the 800 nm lines in the substrate were parallel to the hydrophilic channels and orthogonal to the hydrophobic ones (Fig.8.13a). The silicon substrate was cut by laser ablation at the entrance hole and around the device contour, and then it was bonded to another plain silicon substrate to obtain the master shown in Fig.8.13b. In this way, the PDMS stamp obtained from this master (Fig. 8.13c and Fig.8.13d) has the

8. Applications for 3D Structured Polymerparts

proper thickness and the structure to achieve the entrance through-hole in the final replication. The replication of the PDMS stamp (Fig.8.13e) was performed using the same mixture of UV curing materials and exposure dose described in Chapter 7. The complete opening of the through-hole was easily achieved during the device demolding. A photograph of the released device is shown in Fig.8.14b. SEM images from different

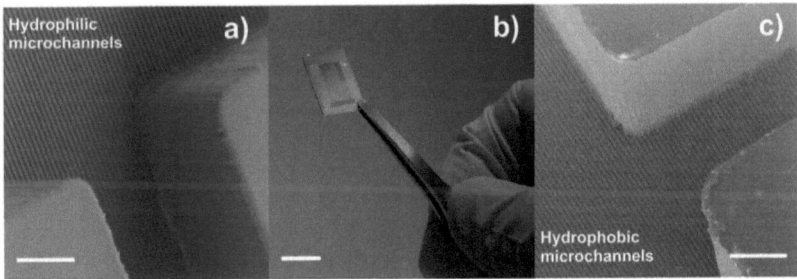

Figure 8.14: Photograph of the replicated microfluidic device (b) and SEM images of the different kind of microchannels in the structure (a and c). Scale bars in B correspond to 1cm and in A and C to 50 μm.

sections of the device can be seen in Fig.8.14a and Fig.8.14c, showing the hydrophilic and hydrophobic channels with the 800 nm lines on their bottom surface. The accuracy of the replication was again satisfactory for all patterns in the device at the micro and nanoscale.

8.2.2.2 Fluidic Characterization and Discussion

The replicated microfluidic devices were bonded to a structure made of SU-8 resin using the method proposed in [159] by applying temperature and pressure. This created closed channels for the fluidic characterization of the device and also proved the compatibility of both materials for further integration in a micro fuel cell. The measurements were performed under a microscope using a 2 molar solution of methanol. In order to compare the effect of the surface modification on the fluidic operation, a fuel delivery system without nanostructures in the surface was also fabricated. To test the devices, a drop of the liquid solution was put on the device inlet. In the case of the device without structures at the bottom, all channels were filled with the liquid (Fig.8.15 a). For the device with the structures at the bottom it was observed that

101

8.2 Fuel Delivery System for Micro Fuel Cell Applications with Passive Feed

the liquid solution flowed through the hydrophilic channels and the hydrophilic area in the middle of the device. As the wetting properties of the channels with orthogonal lines strongly differ from those of the channels with parallel lines, the solution filled the entire hydrophilic area but stopped at the entrance of the hydrophobic channels. Fig.8.15 b shows a picture of one corner of the hydrophilic area and some of

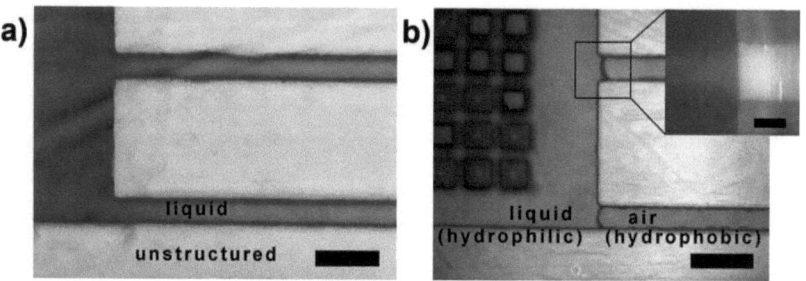

Figure 8.15: Detail view of the replicated devices bonded to a SU-8 cover to close the microchannels, A) Fluidic system without structures in the surface and B) system with 800 nm lines and spaces on the surface (scale bar = 200 μm). The inset shows a detail of the entrance of a hydrophobic channel (scale bar = 50 μm).

the hydrophobic channels after supplying the solution. It can be seen that the fluid is contained in the desired area. The insert in the same figure shows a magnification of one of the hydrophobic channels focused on the surface with nanostructures. The fluid profile creates a sharp edge in line with the hydrophobic structures. In this case, the hydrophobic effect is sufficient to withstand the advance of the fluid. This test was performed several times to evaluate the repeatability of the system, leaving the liquid in the device until it was evaporated and then filling it again. The results of these tests have shown that the function of the device was stable over time. However, in order to achieve a fully working system for the fuel cell application, further optimization is still needed. The hydrophobic effect in these channels was not sufficient to withstand the pressure formed by a CO_2 bubble in the micro fuel cell. An optimized version was fabricated using Thermoforming to increase the hydrophobic effect by applying the structures to the side walls of the channels as well.

8. Applications for 3D Structured Polymerparts

8.2.3 Second Version of the Fuel Delivery System

The second version of the fuel delivery system was fabricated using the Thermoforming process and the foil structuring was performed with Hot Embossing allowing the structures to be applied also to the sidewalls of the channels. The stamp for the Hot Embossing process consists of the same lines and spaces used in the previous section. The structuring of the thermoplastic foil was performed as described in section 6.3.

Figure 8.16: Mold insert for the Thermoforming process fabricated by micro milling of a brass blank.

A mold insert was fabricated by micro milling in a brass blank. A picture of the mold insert is shown in Fig.8.16. The structure consists of 9 micro channels with a width of 200 μm and a depth of 300 μm which are connected on one side to a through hole for fuel contribution and on the other side the channels end on the active area of the fuel cell. The active area in the middle is connected on two sides to 6 micro channels on each side with the same width and depth than the other channels. These channels are foreseen for CO_2 exhaustion. Therefore these channels need hydrophobic wetting properties to retain the fuel and to facilitate the exhaustion of the CO_2 while the channels for fuel contribution need hydrophilic wetting properties. In order to obtain these properties in the different channels, the lines and spaces structure had to be aligned the same way as described before. The hydrophilic channel were aligned parallel while the hydrophobic channels were aligned perpendicular to the lines. As described in the previous sections, the structured foil was further used for the fabrication of a PDMS stamp for the Replica Molding process. A photo of the replicated device is shown in Fig.8.17 A. The iridescence caused by the submicrometer structures can clearly be noticed. For testing the device, the SU 8 cap containing holes in the middle

8.2 Fuel Delivery System for Micro Fuel Cell Applications with Passive Feed

area (Fig.8.17) was bonded on the micro fluidic system. Then a drop of a 2 molar solution of methanol was put on the entrance of the device. The hydrophilic channels transport the solution into the middle area. The hydrophobic channels, which are connected to this area prevent a filling of the channels (Fig.8.17C).

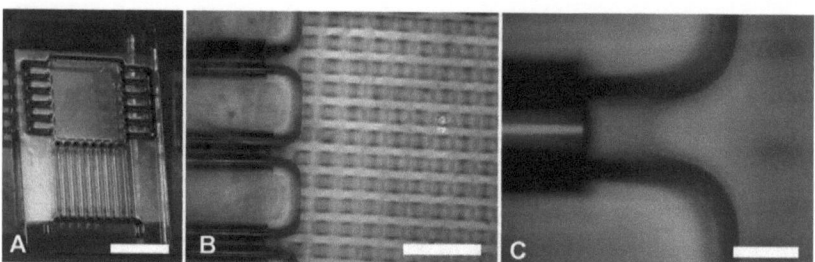

Figure 8.17: Photo of the fuel delivery system using Thermoforming (A) and microscope images of the functional test (B and C)

Even though the hydrophobic effect could be increased by the additional structuring of the side walls, further improvement for the application in the micro fuel cell was still needed, as the channels were filled again, caused by the pressure increment of the produced CO_2 bubbles in the micro fuel cell. The lines and spaces structure only lead to a relative low hydrophobic effect. In order to improve the effect, different structures - for example pillar structures which lead to a stronger hydrophobic effect - can be used. Besides the surface properties of the channels, the design of the fuel delivery system as a whole had to be improved as well to obtain a functional device.

8.2.4 Final Version of the Fuel Delivery System

In the previous section, it was mentioned that a higher hydrophobic effect can be achieved by improving the nanostructures. In the final version of the fuel delivery system the results from the previous fabricated devices were compiled in order to obtain a fully working device. The improvement of the hydrophobic effect was achieved by using new pillar shaped structures with higher aspect ratios. In order to replicate these structures and achieve sufficient stability in the Thermoforming process, the foil structuring was performed by UV-NIL using the method described in section 6.4. Furthermore, the influence of an additional chemical treatment was investigated to improve the hydrophobic effect of the structured surface. The surface properties were also characterized regarding their contact angle using the equipment described in the previous section.

8.2.4.1 Chemical Treatment of the Structured Surfaces and Characterization

The high aspect ratio nanostructures were used to fabricate hydrophobic surfaces. The hydrophobicity of the surface can be increased using an additional chemical treatment. For this purpose a technique was used which is described in [157]. A 50 nm thick gold layer was sputtered on the structured surface, which in turn, was treated for 30 s with oxygen plasma to remove all organic residuals.

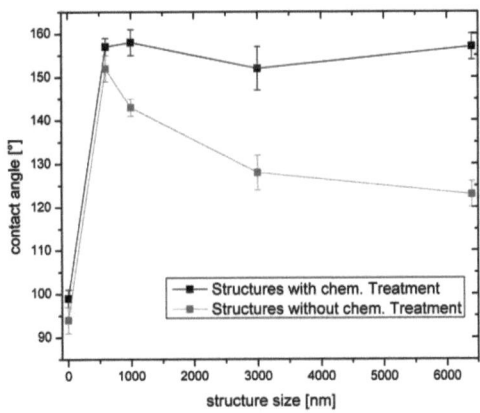

Figure 8.18: Results of the contact angle measurement for samples with and without chemical treatment for various pillar sizes using a 2 molar solution of methanol.

A self-assembled monolayer was deposited by dip coating of octadecanethiol (ODT Merck, Germany). The dip coating of the foils was done in a 0.5 M solution for 10

8.2 Fuel Delivery System for Micro Fuel Cell Applications with Passive Feed

min. After coating, the substrates were dried in a convection oven at 70°C for 10 min. To guarantee the formation of a monolayer of ODT, the substrates were cleaned with isopropanol for 5 min in an ultrasonic bath. The hydrophobic effect of the structured sample and the sample with additional chemical treatment were characterized by measuring the contact angle of the surface. The results of this measurement is illustrated in Fig.8.18 where the mean value of the measured equilibrium angle is plotted. It can be seen that the unstructured substrates result in relatively low contact angle of 94° for the untreated sample and 99° for the chemically treated substrate. Due to the structuring, the contact angle increases significantly to values of 152° for pillars with a diameter of 600 nm and a height of 7 μm without chemical treatment. An increasing diameter of the pillars results in smaller contact angles; a value of 124° for the pillars with a diameter of 6.4 μm was reached. For the chemically treated sample, the contact angle is less affected by the dimensions of the pillars.

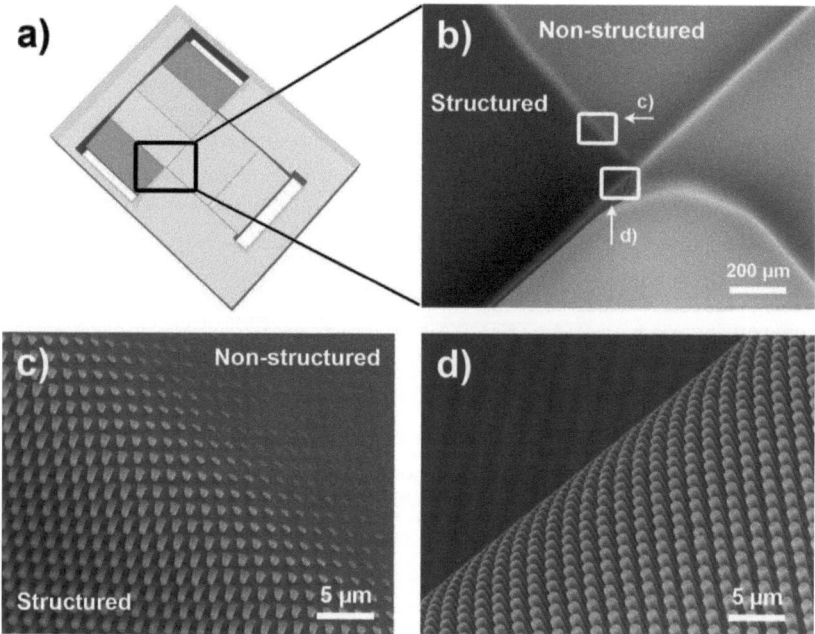

Figure 8.19: SEM images of the fuel delivery component at different magnifications showing the microstructured areas.

The maximum contact angle that can be reached is 158° for the pillars with a diameter of 1 μm. The results of the contact angle measurements have shown that, due to the use of pillar structures with high aspect ratios, higher contact angels can be obtained compared to the previously used lines and spaces structures. Besides the use of the pillar structures to improve the hydrophobic effect of the surface, the design of the

8. Applications for 3D Structured Polymerparts

fluidic system was changed as well. The design of the new device and characteristic details of the device are depicted in Fig.8.19. In this new design the small channel structure has been changed into one big channel for fuel distribution; the channel is connected to a fuel inlet and the active area of the fuel cell. On the sides of the device, the channel structures were also changed into one big channel on each side. These channels are structured with the hydrophobic structures in order to retain the fuel in the active area of the fuel delivery system. Furthermore, the channels have inclinations beginning in the middle of the active area and leading to the slots on the side for CO_2 exhaustion. The inclinations of the chamber and the channels facilitate the exhaustion of the produced CO_2.

8.2.4.2 Fluidic Characterization

In order to characterize the fluidic function, the device was bonded to a SU-8 cap as described before. The test was performed using a 2 molar solution of methanol. In

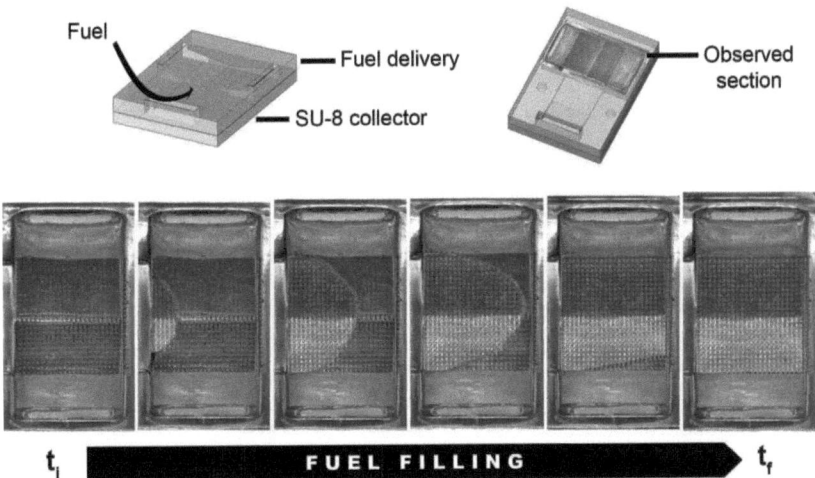

Figure 8.20: Filling process of the fuel delivery process. The fuel delivery system was bonded to a SU-8 current collector to obtain a closed system (top right). The observed area can be seen on the schematic picture.

Fig.8.20, a sequence of pictures of the filling process is depicted. It was observed that the liquid fuel filled the active area of the fuel cell and was retained at the structured areas at the sides. After these tests, the operation of the fuel delivery system in the fuel cell was evaluated. For this evaluation, a hybrid methanol fuel cell, described in the previous section was used.

107

8.2 Fuel Delivery System for Micro Fuel Cell Applications with Passive Feed

Figure 8.21: Assembly of the fuel delivery system placed on top of a hybrid micro direct methanol fuel cell.

The fuel delivery system was placed on top of the anode current collector and the whole assembly was pressed between methacrylate pieces as shown in Fig.8.21. In the previous test, it was demonstrated that the methanol solution only filled the active area and stopped at the hydrophobic surface of the CO_2 outlets. The system was fed with 100 μl of a 2 molar solution of methanol. The fuel cell current output was fixed to 10 mA and the voltage was recorded until the methanol was consumed.

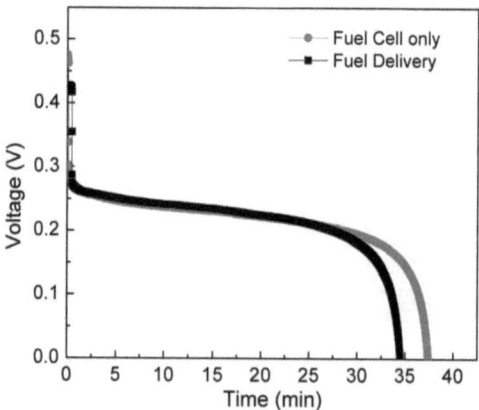

Figure 8.22: Operation of a hybrid micro direct methanol fuel cell operating at 10 mA with and without fuel delivery system

Fig.8.22 shows the operation of the system compared to the response of the hybrid micro fuel cell without the fuel delivery component. It can be seen that the performance of the fuel cell using the fuel delivery component is comparable to the hybrid micro fuel cell

8. Applications for 3D Structured Polymerparts

alone. This test affirms that the fuel delivery system is capable of distributing the fuel to the active area of the fuel cell and remove the produced CO_2 through the integrated hydrophobic gas outlets without compromising the performance of the fuel cell. The use of the fuel delivery system in the fuel cell allows the integration of this power supply system in applications and based on this concept also stacks of fuel cells can be built to obtain higher power. It was shown that the combination of the Thermoforming process together with the Replica Molding process can be used to fabricate a functionalized micro fluidic device which can be incorporated into a micro fuel cell working as a fuel delivery and exhaustion system in a passive way. The presented fabrication technique offers the advantage of altering the surface properties by modifying the topography of the surface allowing the reduction of the device dimensions, material consumption and production costs.

9 Summary and Outlook

9.1 Summary

The main goal of this work was to develop a method for the integration of functional surface structures into 3D micro structured polymer parts to fabricate functionalized polymer devices. For this purpose, structuring techniques capable of replicating submicro and nanometer structures are needed. Therefore, Nanoimprint techniques were developed and applied to the replication of both, functional structures and the structuring of functional materials.

In the first step the HE-NIL process was developed, which was applied to the fabrication of photonic crystals. The characterization of the photonic crystals has shown that after process optimization, high quality photonic crystals could be replicated. The challenge of the replication of the photonic crystals, using imprint technologies, is that the design contains both waveguide structures in the range of various micrometers and the photonic crystal structure itself with dimensions in the sub 100 nm range. Furthermore, in the case of photonic crystals, small failures in the structure attributed to replication errors lead to extreme failures in the function of the photonic crystal. The comparison of the measurement results with the simulation have shown that the replicated photonic crystals are of high quality. The HE-NIL process was further used for the structuring of functional materials for applications in solar cells as well as for magnetic applications. The results from these experiments have shown that this technology is also suitable for the structuring of different materials in the sub 100 nm range.

In order to integrate functional structures into micro or macrosystems, a Thermoforming process was used. For this Thermoforming process, the experience gained in the first NIL experiments was used to create functional structures on a thermoplastic foil. The structuring of the thermoplastic foil was done using both Hot Embossing and UV-NIL. The structured foils were then used in the Thermoforming process. The development of the Thermoforming process required the design and the fabrication of a Thermoforming tool and a pressure system. The Thermoforming tool was used to cover a microstructure with the structured thermoplastic foil and, then, to fabricate a PDMS stamp containing the functionalized microstructres. In the tool, different mold inserts containing the microstructures can be used. The mold inserts can be fabricated using different techniques, for example, micro milling. In the first approach, the foil structuring was done using Hot Embossing. In this case, the functional structures are imprinted directly into the thermoplastic foil. Therefore, the parameters (temperature

9.1 Summary

and pressure) during the Thermoforming process had to be optimized carefully. A high pressure and temperature led to a high structure fidelity of the microstructures but to a poor quality of the functional structures as they are made out of the same material and are therefore softened as well. Furthermore, in the first approach the mold inserts, fabricated using micro milling, had to be electro-polished to eliminate the burr at the corners of the mold insert. The burr and the sharp corners of the mold insert are critical, because they can damage the foil due to local increase of stress.

In the second approach, these critical points were optimized by using a new foil assembly, specially developed to overcome these limitations. In order to obtain a higher stability of the foil during the Thermoforming process, the backside of the foil was prepared with a thin PDMS membrane. Therefore, the necessary electro polishing step could be prevented without damaging the foil. Furthermore, the foil structuring in the second approach was done using UV-NIL. Due to the UV-NIL process, two restrictions were eliminated. First, the functional structures are fabricated in a material with a higher temperature stability and secondly higher aspect ratios can be obtain in this process. The higher temperature stability allowed the use of higher temperatures and pressures in the Thermoforming process without affecting the quality of the functional structures. Due to this optimization, a higher structure fidelity of both the microstructures and the functional structures could be obtained.

The fabrication of the PDMS stamp could be obtained in the Thermoforming tool by integration of a ring on top of the structured foil. The PDMS stamp could be fabricated without removing the structured foil from the tool and then applied to a Replica Molding process for the fabrication of 3D structured polymer parts as well as to a UV-NIL process for the fabrication of 3D structured polymer films. For the replication process, epoxy based UV curable materials were used. These materials were cast onto the PDMS stamp and cured by UV exposure to fabricate 3D structured polymer parts. For the UV-NIL process a layer of the epoxy material was coated onto a substrate, brought into contact with the PDMS stamp and also cured by UV exposure. The viscosity of the epoxy material was adjusted to the requirements of the process by changing the composition of the mixture and the process temperature. In this way it is possible to use these material systems for the different processes.

The combination of the Thermoforming process and the Soft Lithography process (Replica Molding and UV-NIL with PDMS stamps) was used for the actualization of different applications. Micro and nanostructures have influence on various surface properties. In this work the process chain was used for the functionalization of micro lenses and a fluidic system using two different physical effects. Micro lenses form an important part of many engineering devices such as optical sensors, LEDs and camera modules. The optical function of micro lenses can be improved by applying a periodic surface relief on subwavelength scale on the surface. These so called moth eye structures

9. Summary and Outlook

result in a gradual change of the refractive indices from one medium to another. In this work a master containing these moth eye structure was obtained from the Fraunhofer Institute for Solar Energy in Freiburg. The moth eye structures have a period of 280 nm and a a height of about 200 nm. The structures were fabricated using Interference Lithography. Using Interference Lithography offers the advantage that the structures can be applied to large areas. A disadvantage of this method is that there are some limitations regarding the minimal period and maximum height of the structures. In order to obtain an optimal anti-reflective effect for applications in the visible range, it is known from literature that the height of the structures should be at least 250 nm and the period should be as small as possible. However, the structures used in this work have shown good anti-reflective properties in experiments. The master for the lens array was produced using Contactless Embossing of Microlenses (CEM). The master containing the micro lenses was covered with a foil containing the moth eye structures. The produced PDMS stamp was applied to a UV-NIL process allowing the replication of functionalized micro lenses. The characterization was done by measuring the transmission of a sample containing the moth eye structures and compared to the spectra taken from an unstructured sample. The results have shown that a significant higher transmission could be measured for the structured sample. Therefore, this method can be used for the fabrication and replication of functionalized micro lenses.

The second application actualized is the fabrication of a functionalized micro fluidic system. The fluidic system was used as a fuel delivery and out-gassing system for a micro direct methanol fuel cell. These functionalities were obtained by integrating hydrophilic and hydrophobic areas into the fluidic system. The hydrophobicity was obtained using pillar structures. Contact angle measurements have shown that a higher contact angle could be obtained with pillars having a high aspect ratio. In order to integrate hydrophobic areas, a pillar structure with an aspect ratio of 8 ($1\mu m$ in diameter) was used. The fluidic system was designed in a way that the liquid fuel can pass through the hydrophilic (unstructured) area into the fuel cell and stops at the hydrophilic-hydrophobic (structured) junction. The gas by-products produced during the operation of the fuel cell can escape through the hydrophobic areas and the integrated through hole. The fluidic devices were fabricated using the produced PDMS stamp in a Replica Molding process allowing the fabrication of functionalized polymer parts. The characterization of the fluidic function was done by integrating the fluidic system into a fuel cell developed at the Centro Nacional de Microelectronica (CNM) in Barcelona. Various fluidic designs were tested and optimizations were actualized. The results of the final system have shown that the operation of the fuel cell was not affected by the integration of the fluidic system. The correct function of the fuel cell demonstrated that both objectives (fuel delivery and out-gassing) were obtained.

9.2 Outlook

The results of this work have shown the potential of NIL and related, newly developed process variations for different applications. It has been shown that nanostructuring based on imprint technologies can be useful for many applications in fields like information, energy or sensor technology. Further improvement, especially for parameters like repeatability and defect control, is needed to make NIL based processes usable as industrial production processes. As demonstrated in this work a reliable nanostructuring process can lead to the development of completely new products like improved solar cells, miniaturized magnetic sensors and new applications in laser or sensor technology based on photonic crystals.

With the newly developed process chain, combining Thermoforming and Replica Molding for the fabrication of 3D structured polymer parts very promising results were obtained. It is a very stable process which offers the possibility of fabricating and replicating functional structures on 3D surfaces with high aspect ratios and a good resolution. It was demonstrated that with this method sidewall patterning, which is especially interesting for microfluidics, and the patterning of 3D surfaces, which is interesting for optics, can be obtained.

The possibility of scaling up the entire process and the low capital investment makes this process attractive for industrial applications and offers the possibility of designing new product with specific functionalization.

Abbreviations

NIL	-	Nanoimprint Lithography
EBL	-	Electron Beam Lithography
UV-NIL	-	Ultra Violet Nanoimprint Lithography
HE-NIL	-	Hot Embossing Nanoimprint Lithography
NGL	-	Next Generation Lithography
2D	-	twodimensional
3D	-	threedimensional
EUV	-	Extreme Ultra Violet
RIE	-	Reactive Ion Etching
PMMA	-	poly(methyl methacrylate)
SFIL	-	Step and Flash Imprint Lithography
EPS	-	Element Patterned Stamp
CNP	-	Combined Nanoimprint Photolithography
PDMS	-	poly(dimethylsiloxane)
BCE	-	Bis-cycloaliphatic epoxide
DGE-TEG	-	Divinylether of triethylene glycol
DGE-BPA	-	Diglycidylether of bis-phenol A
TAS	-	Triarylsulfonium salt
PP	-	Polypropylene
PS	-	Polystyrene
PC	-	Polycarbonate
SMART	-	Surface Modification and Replication by Thermoforming
SEM	-	Scanning Electron Microscope
MEMS	-	Microelectromechanical System
DMF	-	Dimethylformamid
ICP	-	Inductive Coupled Plasma
RF	-	Radio Frequency
F13-TCS	-	(Tridecafluoro-1,1,2,2,-tetrahydrooctyl)-trichlorsilane
XPS	-	X-Ray Photoemission Spectroscopy
VCSEL	-	Vertical Surface Emitting Laser
PBG	-	Photonic Band Gap
FDTD	-	Finite Difference Time Domain
LPCVD	-	Low Pressure Chemical Vapour Deposition
MOVPE	-	Metalorganic Vapour Phase Epitaxy
CNT	-	Carbon Nano Tubes
MOKE	-	Magneto Optical Kerr Effect
SQUID	-	Superconductive quantum interference device

GMR	-	Giant Magnetoresistance
LED	-	Light Emitting Diode
CEM	-	Contactless Embossing of Microlenses
MEA	-	Membrane Electrode Assembly
CNM	-	Centro National de Microelectronica
ODT	-	Octadecanthiol

Formula Symbols

T_g	-	Glass Transition Temperature [°C]
$T_{imprint}$	-	Imprint Temperature [°C]
M_W	-	Molecular Weight [kg/mol]
T_{demold}	-	Demold Temperature [°C]
η	-	Viscosity [Pas]
L	-	Length of a Cavity [m]
h_r	-	Height of a Cavity [m]
s_i	-	Distance of a Cavity [m]
w_i	-	Width of a Cavity [m]
h_f	-	Residual Layer Thickness [m]
h_0	-	Initial Layer Thickness [m]
ν	-	Fill Factor
h(t)	-	Film Thickness [m]
F	-	Force [N]
p	-	Pressure [MPa]
R_a	-	Areal Draw Ratio
$A_{sidewall}$	-	Area of the Sidewall of a Cylindrical Mold [m²]
A_{bottom}	-	Area of the Bottom of a Cylindrical Mold [m²]
A_0	-	Projected Area of the Opening of a Cylindrical Mold [m²]
R_L	-	Linear Draw Ratio [m]
s	-	Height of a Cylindrical Mold [m]
d	-	Diameter of a Cylindrical Mold [m]
a	-	Lattice Constant of a Photonic Crystal [m]
d	-	Radius of the Holes of a Photonic Crystal [m]
t	-	Thickness of the Dielectric Slab [m]
T_{thermo}	-	Thermoforming Temperature [°C]
p_{thermo}	-	Thermoforming Pressure [bar]
$p_{imprint}$	-	Imprint Pressure [MPa]
D_1	-	Distance between two Pillars at a Bottom Corner [m]
D_{0p}	-	Original Distance of the Heads of the Pillars [m]
h	-	Height of the Pillars [m]
r	-	Curvature Radius [m]
t	-	Thickness of the Foil including the PDMS Membrane [m]

List of Tables

2.1 Comparison of HE-NIL and UV-NIL with typical parameters (in modified form from [10]) 15

5.1 Process requirements for the replication of photonic crystals using HE-NIL 44

5.2 The structure height for the optimized imprints for various stamp depths 46

6.1 Requirements for the tool and pressure system for the Thermoforming process. 56

List of Figures

2.1 Schematic overview over the process steps during NIL 7

2.2 Temperature and pressure sequence during the HE-NIL process[10] . . 8

2.3 Schematic overview for the resist patterning using the CNP process. . . 9

2.4 Influence of the molecular weight (M_W) and temperature on the storage modulus of a thermoplastic polymer material[10]. 12

2.5 Schematic image of a stamp and a coated substrate before and after the imprint [23]. 16

2.6 Filling sequence of a single cavity during an imprint process [23]. 18

2.7 Different types of Thermoforming.[57] 19

2.8 Typical thickness variation of a foil during the Thermoforming process[57] 20

2.9 Examples of 3D structures obtained using Thermoforming with pre-processed foils[42, 68] 22

2.10 Reaction equation for the polymerization of PDMS 24

2.11 Defect mechanism in Soft Lithography methods. 25

2.12 SEM pictures of the defect mechanism of PDMS. 26

3.1 Schematic overview over the process steps for the fabrication of imprint stamps. 30

3.2 Results from the fabrication process for the imprint stamps. 31

3.3 Bonding of F13-TCS on Si as passivation of imprint stamps[42]. 32

3.4 Process steps for the fabrication of a PDMS stamp and the replication of a pattern using UV-NIL. 34

4.1 SEM images of the results of the HE-NIL process. 36

4.2 Influence of the holding time on the residual layer for a small structure detail. 37

LIST OF FIGURES

4.3 Influence of the holding time on the residual layer for a bigger structure detail. 38

5.1 Design of the photonic crystal test structure. 41

5.2 Schematic image of a photonic crystal filter structure 42

5.3 Schematic image of a L3 cavity 42

5.4 Schematic overview over the process steps during NIL for the replication of photonic crystals. 43

5.5 SEM image of an uncompleted nitride layer etching. 45

5.6 Sectional view of an imprinted structure and structure height in dependence of the process time. 46

5.7 SEM picture of replicated photonic crystal structure and corresponding stamp. 47

5.8 Transmission spectra of the photonic crystal structure and the corresponding 3D FDTD simulation. 48

5.9 SEM images of the controlled growth of CuInSe compounds on a pre-structured Si substrate. 49

5.10 SEM pictures of structured magnetic layers. 50

5.11 Test structure for the measurement of the influence of nanostructuring on the GMR effect. 51

5.12 Fabricaion process for the test structure to measure the influence of nanostructuring on the GMR effect. 51

5.13 Schematic and SEM image of the layer assembly after NIL. 52

5.14 Schematic and SEM image of the layer assembly after Lift Off. 52

6.1 Schematic overview over the pressure system used for the Thermoforming process. 57

6.2 Schematic drawing of the Thermoforming tool and a micro milled brass mold insert. 58

6.3 Result of the Thermoforming process for an unstructured foil. 58

6.4 3D structuring of a thermoplastic foil. 60

LIST OF FIGURES

6.5 Comparison of the temperature pressure sequence during Hot Embossing and Thermoforming. 61

6.6 Replication results from the Thermoforming process. 62

6.7 Process overview for the fabrication of 3D structured foils using UV-NIL. 64

6.8 Temperature-dependent viscosity of epoxy material and thickness of PDMS membranes at different spinning velocities 65

6.9 SEM image of a cross-sectional view of a thermoformed microchannel. . 67

6.10 SEM images of the entrance of a micro channel. 67

6.11 Result of the Thermoforming process for high aspect ratio nanostructures. 68

6.12 Cross sectional view of a thermoformed foil containing high aspect ratio structures. 69

6.13 Foil thickness distribution of a multilayer foil after thermoforming. . . . 70

6.14 Schematic of the couette flow in the foil[129]. 71

6.15 Influence of temperature and pressure on the Thermoforming result. . . 72

6.16 Schematic image of the foil deformation after Thermoforming process and a SEM picture which visualize the effect of foil deformation. 73

6.17 Dependence between foil deformation and pillar distance 74

6.18 SEM pictures of the edge of a thermoformed foil. 74

6.19 Thermoforming result for different foil thickness on a pillar field I. . . . 75

6.20 Thermoforming result for different foil thickness on a pillar field II. . . 76

7.1 Schematic diagram of the Replica Molding process. 80

7.2 SEM images of the master structure and the replications for nanostructure replication. 82

7.3 SEM images of the process for multilevel replication 83

7.4 Replica molding process for 3D polymer replications. 84

7.5 SEM images illustrating the process steps for 3D polymer replications. 85

7.6 SEM pictures of a replicated structure. 85

LIST OF FIGURES

7.7 Schematic image of the demolding mechanism during Replica Molding.   86

7.8 SEM picture of a cross sectional view of a replication containing high aspect ratio nanostructures. 87

8.1 Typical fabrication process for optical modules (REEMO®)[138]. . . . 89

8.2 Schematic diagram for the integration of the moth eye structures into the PDMS stamp and the replication of the functionalized lens array. . 91

8.3 Master structure for the fabrication of a lens array and replicated lens array. 92

8.4 Results from the UV/VIS spectroscopy for a structured and a unstructured replication. 93

8.5 Structured foil, PDMS stamp and replication in epoxy containing the moth eye structures 93

8.6 Digital microscope and SEM image of structured foil containing the moth eye structures. 94

8.7 Digital microscope and SEM image of the replicated lens array containing the moth eye structures. 94

8.8 Results of the AFM measurement of the original master structure and the replicated lens array. 95

8.9 Hybrid fuel cell parts and assembly [153]. 97

8.10 Filling of the anode reservoir of the micro fuel cell [153]. 98

8.11 Basic design of the fuel delivery system, integration into the fuel cell and function [153]. 98

8.12 Contact angle measurement values obtained from a line-structured surface. 99

8.13 Fabrication process of the fuel delivery system. 100

8.14 Photograph of the replicated microfluidic device. 101

8.15 Detail view of the replicated devices bonded to a SU-8 cover 102

8.16 Mold insert for the Thermoforming process. 103

8.17 Photo of the fuel delivery system using the Thermoforming process . . 104

LIST OF FIGURES

8.18 Results of the contact angle measurement for samples with and without chemical treatment for various pillar sizes. 105

8.19 SEM images of the fuel delivery component at different magnifications. 106

8.20 Filling process of the fuel delivery process. 107

8.21 Assembly of the fuel delivery system placed on top of a hybrid micro direct methanol fuel cell. 108

8.22 Operation of a hybrid micro direct methanol fuel cell 108

References

[1] B. Bhushan. *Handbook of Nanotechnology 3rd edition, Preface.* Springer-Verlag, 2010.

[2] E. Yablonovitch. Inhibited spontaneous emission in Solid-State physics and electronics. *Physical Review Letters*, 58(20):2059, May 1987.

[3] D. S. Hobbs, B. D. MacLeod, and J. R. Riccobono. Update on the development of high performance anti-reflecting surface relief micro-structures. In *Proceedings of SPIE*, pages 65450Y–65450Y–14, Orlando, FL, USA, 2007.

[4] A. Gombert, B. Bläsi, Ch. Bühler, P. Nitz, J. Mick, W. Hoßfeld, and M. Niggemann. Some application cases and related manufacturing techniques for optically functional microstructures on large areas. *Optical Engineering*, 43(11):2525, 2004. ISSN 00913286.

[5] W. Barthlott and C. Neinhuis. Purity of the sacred lotus, or escape from contamination in biological surfaces. *Planta*, 202(1):1–8, April 1997. ISSN 0032-0935.

[6] S J Abbott and P H Gaskell. Mass production of bio-inspired structured surfaces. *Proceedings of the Institution of Mechanical Engineers, Part C: Journal of Mechanical Engineering Science*, 221(10):1181–1191, January 2007. ISSN 0954-4062.

[7] S. Y. Chou, P. R. Krauss, and P. J. Renstrom. Imprint of sub-25 nm vias and trenches in polymers. *Applied Physics Letters*, 67(21):3114, 1995. ISSN 00036951.

[8] L. J. Guo. Nanoimprint lithography: Methods and material requirements. *Advanced Materials*, 19:495–513, 2007.

[9] Y. Xia, E. Kim, X.-M. Zhao, J. A. Rogers, M. Prentiss, and G. M. Whitesides. Complex optical surfaces formed by replica molding against elastomeric masters. *Science*, 273(5273):347–349, July 1996.

[10] H. Schift and A. Kristensen. *Handbook of Nanotechnology 3rd edition, chapter 9.* Springer-Verlag, 2010.

[11] P. B. Clapham and M. C. Hutley. Reduction of lens reflexion by the "moth eye" principle. *Nature*, 244(5414):281–282, 1973.

[12] A. Schleunitz and H. Schift. Fabrication of 3D patterns with vertical and sloped sidewalls by grayscale electron-beam lithography and thermal annealing. *Microelectronic Engineering*, 88(8):2736–2739, August 2011. ISSN 0167-9317.

REFERENCES

[13] A. Schleunitz and H. Schift. Fabrication of 3D nanoimprint stamps with continuous reliefs using dose-modulated electron beam lithography and thermal reflow. *Journal of Micromechanics and Microengineering*, 20(9):095002, September 2010. ISSN 0960-1317.

[14] A. Gombert, W. Glaubitt, K. Rose, J. Dreibholz, B. Bläsi, A. Heinzel, D. Sporn, W. Döll, and V. Wittwer. Subwavelength-structured antireflective surfaces on glass. *Thin Solid Films*, 351(1-2):73–78, August 1999. ISSN 0040-6090.

[15] S. Möllenbeck, N. Bogdanski, M. Wissen, H.-C. Scheer, J. Zajadacz, and K. Zimmer. Investigation of the separation of 3D-structures with undercuts. *Microelectronic Engineering*, 84(5-8):1007–1010, May 2007. ISSN 0167-9317.

[16] S. Möllenbeck, N. Bogdanski, M. Wissen, H.-C. Scheer, J. Zajadacz, and K. Zimmer. Multiple replication of three dimensional structures with undercuts. *Journal of Vacuum Science & Technology B: Microelectronics and Nanometer Structures*, 25(1):247, 2007. ISSN 10711023.

[17] S. Giselbrecht, T. Gietzelt, E. Gottwald, C. Trautmann, R. Truckenmüller, K. F. Weibezahn, and A. Welle. 3D tissue culture substrates produced by microthermoforming of pre-processed polymer films. *Biomedical Microdevices*, 8(3):191–199, 2006. ISSN 1387-2176.

[18] S. Y. Chou and P. R. Krauss. Imprint lithography with sub-10 nm feature size and high throughput. *Microelectronic Engineering*, 35(1-4):237–240, February 1997. ISSN 0167-9317.

[19] M. D. Austin, H. Ge, W. Wu, M. Li, Z. Yu, D. Wasserman, S. A. Lyon, and S. Y. Chou. Fabrication of 5nm linewidth and 14nm pitch features by nanoimprint lithography. *Applied Physics Letters*, 84(26):5299, 2004. ISSN 00036951.

[20] S. Y. Chou. Nanoimprint lithography. *Journal of Vacuum Science & Technology B: Microelectronics and Nanometer Structures*, 14(6):4129, 1996. ISSN 0734211X.

[21] S. Y. Chou. Sub-10 nm imprint lithography and applications. *Journal of Vacuum Science & Technology B: Microelectronics and Nanometer Structures*, 15(6):2897, 1997. ISSN 0734211X.

[22] The International Technology Roadmap for Semiconductors website. http://public.itrs.net/, accessed March 25 2011.

[23] C. M. Sotomayor Torres. *Alternative Lithography*. Kluwer Academic Publishers, 2003.

[24] A. del Campo and E. Arzt. Fabrication approaches for generating complex micro- and nanopatterns on polymeric surfaces. *Chem. Rev.*, 108:911–945, 2008.

REFERENCES

[25] A. Schleunitz, T. Senn, D. Proyag, J. Göttert, S. Giselbrecht, M. Reinhardt, and B. Löchel. Nanoprägen in Polymerfolien als Fertigungsverfahren für fluidische Mikro-Nano-Systeme. In *MST Kongress, Berlin*, 2009.

[26] A. Schleunitz, C. Spreu, T. Mäkelä, T. Haatainen, A. Klukowska, and H. Schift. Hybrid working stamps for high speed roll-to-roll nanoreplication with molded sol-gel relief on a metal backbone. *Microelectronic Engineering*, 88(8):2113–2116, August 2011. ISSN 0167-9317.

[27] J. Haisma. Mold-assisted nanolithography: A process for reliable pattern replication. *Journal of Vacuum Science & Technology B: Microelectronics and Nanometer Structures*, 14(6):4124, November 1996. ISSN 0734211X.

[28] M. Colburn. Step and flash imprint lithography: a new approach to high-resolution patterning. In *Proceedings of SPIE*, pages 379–389, Santa Clara, CA, USA, 1999.

[29] M. Colburn, T. Bailey, B. Choi, J. Ekerdt, S. Sreenivasan, and C.G. Willson. Development and advantages of step-and-flash lithography. *Solid State Technology*, 44(7), July 2001.

[30] M. Otto, M. Bender, B. Hadam, B. Spangenberg, and H. Kurz. Characterization and application of a UV-based imprint technique. *Microelectronic Engineering*, 57-58:361–366, September 2001. ISSN 0167-9317.

[31] D. J. Resnick, W. J. Dauksher, D. Mancini, K. J. Nordquist, T. C. Bailey, S. Johnson, N. Stacey, J. G. Ekerdt, C. G. Willson, S. V. Sreenivasan, and N. Schumaker. Imprint lithography for integrated circuit fabrication. *Journal of Vacuum Science & Technology B: Microelectronics and Nanometer Structures*, 21(6):2624, 2003. ISSN 0734211X.

[32] J. h. Jeong, Y. s. Sim, H. Sohn, and E. s. Lee. UV-nanoimprint lithography using an elementwise patterned stamp. *Microelectronic Engineering*, 75(2):165–171, August 2004. ISSN 0167-9317.

[33] X. Cheng and L. J. Guo. A combined-nanoimprint-and-photolithography patterning technique. *Microelectronic Engineering*, 71(3-4):277–282, May 2004. ISSN 0167-9317.

[34] Younan Xia and George M. Whitesides. SOFT LITHOGRAPHY. *Annual Review of Materials Science*, 28(1):153–184, 1998. ISSN 0084-6600.

[35] M. Bender, A. Fuchs, U. Plachetka, and H. Kurz. Status and prospects of UV-Nanoimprint technology. *Microelectronic Engineering*, 83(4-9):827–830, April 2006. ISSN 0167-9317.

REFERENCES

[36] S. H. Ahn and L. J. Guo. High-Speed Roll-to-Roll nanoimprint lithography on flexible plastic substrates. *Advanced Materials*, 20(11):2044–2049, 2008. ISSN 09359648.

[37] S. H. Ahn and L. J. Guo. Large-Area Roll-to-Roll and Roll-to-Plate nanoimprint lithography: A step toward High-Throughput application of continuous nanoimprinting. *ACS Nano*, 3(8):2304–2310, 2009.

[38] H. Hauser, P. Berger, B. Michl, C. Mueller, S. Schwarzkopf, M. Hermle, and B. Bläsi. Nanoimprint lithography for solar cell texturisation. *MICRO-OPTICS 2010*, 7716, 2010. ISSN 0277-786X.

[39] M. Bender, M. Otto, B. Hadam, B. Spangenberg, and H. Kurz. Multiple imprinting in UV-based nanoimprint lithography: related material issues. *Microelectronic Engineering*, 61-62:407–413, July 2002. ISSN 0167-9317.

[40] N. Kehagias, V. Reboud, G. Chansin, M. Zelsmann, C. Jeppesen, C. Schuster, M. Kubenz, F. Reuther, G. Gruetzner, and C. M. Sotomayor Torres. Reverse-contact UV nanoimprint lithography for multilayered structure fabrication. *Nanotechnology*, 18(17):175303, 2007. ISSN 0957-4484.

[41] L.-R. Bao, X. Cheng, X. D. Huang, L. J. Guo, S. W. Pang, and A. F. Yee. Nanoimprinting over topography and multilayer three-dimensional printing. *Journal of Vacuum Science & Technology B: Microelectronics and Nanometer Structures*, 20(6):2881, 2002. ISSN 0734211X.

[42] A. Schleunitz. *Untersuchungen zum Nanoprägen als Fertigungsverfahren für fluidische Mikro-Nano-Systeme*. PhD thesis, Universität Karlsruhe, 2009.

[43] Y. Xia, J.J. McClelland, R. Gupta, D. Qin, X.-M. Zhao, L.L. Sohn, R.J. Celotta, and G.M. Whitesides. Replica molding using polymeric materials: A practical step toward nanomanufacturing. *Advanced Materials*, 9:147–149, 1997.

[44] S. C. Johnson. Advances in step and flash imprint lithography. In *Proceedings of SPIE*, pages 197–202, Santa Clara, CA, USA, 2003.

[45] C. Schuster, F. Reuther, A. Kolander, and G. Gruetzner. mr-NIL 6000LT - epoxy-based curing resist for combined thermal and UV nanoimprint lithography below 50 °C. *Microelectronic Engineering*, 86(4-6):722–725, April 2009. ISSN 0167-9317.

[46] Tobias Senn. Entwicklung und Charakterisierung innovativer Resistsysteme für die UV-NIL. Master's thesis, Universität Freiburg, IMTEK, 2008.

[47] T. Senn, C. Mueller, and H. Reinecke. Replication of HARMST and large area nanostructured parts using UV cationic polymerization. *Journal of Micromechanics and Microengineering*, 20(7):075002, 2010. ISSN 0960-1317.

REFERENCES

[48] Y.M.Kim, L.K. Konstanski, and J.F. MacGregor. Photopolymerization of 3,4-epoxycyclohexylmethyl-3,4-epoxycyclohexanecarboxylate and tri(ethylene glycol) methyl vinyl ether. *Polymer*, 44:5103–5109, 2003.

[49] F. Sun, S.L. Jiang, and J. Liu. Study on cationic photopolymerization reaction of epoxy polysiloxane. *Nuclear Instruments and Methods in Physics Research B*, 264:318–322, 2007.

[50] C. Decker, T. Ngyuen Thi Viet, D. Decker, and E. Weber Koehl. Uv-radiation curing of acrylate epoxide systems. *Polymer*, 42:5531–5541, 2001.

[51] C. Decker, C. Bianchi, D. Decker, and F. Morel. Photoinitiated polymerization of vinylether based system. *Progress in Organic Coatings*, 42:253–266, 2001.

[52] S. Wu, M.T. Sears, M.D. Soucek, and W.J. Simonsick. Synthesis of reactive diluents for cationic cycloaliphatic epoxide uv coatings. *Polymer*, 40:5675–5686, 1999.

[53] T.H. Chiang and T.E. Hsieh. A study of monomerŽs effect on adhesion strength of uv-curable resins. *International Journal of Adhesion & Adhesives*, 26:520–531, 2006.

[54] H. Schift. Nanoimprint lithography: An old story in modern times? a review. *Journal of Vacuum Science & Technology B: Microelectronics and Nanometer Structures*, 26(2):458, 2008. ISSN 10711023.

[55] L. J. Heyderman, H. Schift, C. David, J. Gobrecht, and T. Schweizer. Flow behaviour of thin polymer films used for hot embossing lithography. *Microelectronic Engineering*, 54(3-4):229–245, December 2000. ISSN 0167-9317.

[56] H. -C. Scheer and H. Schulz. A contribution to the flow behaviour of thin polymer films during hot embossing lithography. *Microelectronic Engineering*, 56:311–332, August 2001. ISSN 0167-9317. ACM ID: 567720.

[57] J. L. Throne. *Technology of Thermoforming*. Carl Hanser Verlag, München, 1996.

[58] R. Truckenmüller, Z. Rummler, Th. Schaller, and W. K. Schomburg. Low-cost thermoforming of micro fluidic analysis chips. *Journal of Micromechanics and Microengineering*, 12(4):375–379, 2002. ISSN 09601317.

[59] A. Disch, C. Mueller, and H. Reinecke. Low cost production of disposable microfluidics by blister packaging technology. *Conference Proceedings: Annual International Conference of the IEEE Engineering in Medicine and Biology Society. IEEE Engineering in Medicine and Biology Society. Conference*, 2007:6323–6326, 2007. ISSN 1557-170X. PMID: 18003467.

[60] E. Gottwald, S. Giselbrecht, C. Augspurger, B. Lahni, N. Dambrowsky, R. Truckenmuller, V. Piotter, T. Gietzelt, O. Wendt, W. Pfleging, A. Welle, A. Rolletschek, A. M. Wobus, and K.-F. Weibezahn. A chip-based platform for the in vitro generation of tissues in three-dimensional organization. *Lab on a Chip*, 7 (6):777, 2007. ISSN 1473-0197.

[61] R. Truckenmüller, S. Giselbrecht, C. van Blitterswijk, N. Dambrowsky, E. Gottwald, T. Mappes, A. Rolletschek, V. Saile, C. Trautmann, K.F. Weibezahn, and A. Welle. Flexible fluidic microchips based on thermoformed and locally modified thin polymer films. *LAB ON A CHIP*, 8(9):1570–1579, September 2008. ISSN 1473-0197.

[62] M. Focke, D. Kosse, C. Müller, H. Reinecke, R. Zengerle, and F. von Stetten. Lab-on-a-Foil: microfluidics on thin and flexible films. *Lab on a Chip*, 10(11): 1365, 2010. ISSN 1473-0197.

[63] M. Focke, F. Stumpf, G. Roth, R. Zengerle, and F. von Stetten. Centrifugal microfluidic system for primary amplification and secondary real-time PCR. *Lab on a Chip*, 10(23):3210, 2010. ISSN 1473-0197.

[64] C. Richter, M. Reinhardt, S. Giselbrecht, D. Leisen, V. Trouillet, R. Truckenmuller, A. Blau, C. Ziegler, and A. Welle. Spatially controlled cell adhesion on three-dimensional substrates. *BIOMEDICAL MICRODEVICES*, 12(5):787–795, October 2010. ISSN 1387-2176.

[65] M. Heilig, S. Giselbrecht, A. Guber, and M. Worgull. Micro thermoforming of nanostructured polymer films: A new bonding method for the integration of nanostructures in 3-dimensional cavities. In *Design, Test, Integration & Packaging of MEMS/MOEMS, 2009. MEMS/MOEMS '09. Symposium on*, pages 132–135, 2009.

[66] M. Heilig, M. Schneider, H. Dinglreiter, and M. Worgull. Technology of microthermoforming of complex three-dimensional parts with multiscale features. *Microsystem Technologies*, 17(4):593–600, 2011. ISSN 0946-7076.

[67] M. Reinhardt. *Funktionalisierte, polymere Mikrostrukturen für die dreidimensionale Zellkultur*. PhD thesis, Karlsruhe Institut für Technologie, 2010.

[68] Stefan Giselbrecht, Martina Reinhardt, Timo Mappes, Martin Börner, Eric Gottwald, Clemens van Blitterswijk, Volker Saile, and Roman Truckenmüller. Closer to nature-bio-inspired patterns by transforming latent lithographic images. *Advanced Materials*, 2011. ISSN 1521-4095.

[69] R. Truckenmüller, S. Giselbrecht, N. Rivron, E. Gottwald, V. Saile, A. van den Berg, M. Wessling, and C. van Blitterswijk. Thermoforming of Film-Based

REFERENCES

biomedical microdevices. *ADVANCED MATERIALS*, 23(11):1311–1329, March 2011. ISSN 0935-9648.

[70] L. Zimmermann. *Dreidimensional nanostrukturierte und superhydrophobe mikrofluidische Systeme zur Tröpfchengenerierung und - handhabung*. PhD thesis, Karlsruher Institut für Technologie, 2010.

[71] Y. Xia and G.M. Whitesides. Soft lithography. *Angewandte Chemie*, 37:550–575, 1998.

[72] H. Masuda and K. Fukuda. Ordered metal nanohole arrays made by a Two-Step replication of honeycomb structures of anodic alumina. *Science*, 268(5216):1466–1468, June 1995.

[73] P. Hoyer, N. Baba, and H. Masuda. Small quantum-sized CdS particles assembled to form a regularly nanostructured porous film. *Applied Physics Letters*, 66(20):2700, 1995. ISSN 00036951.

[74] D. Y. Kim, S. K. Tripathy, Lian Li, and J. Kumar. Laser-induced holographic surface relief gratings on nonlinear optical polymer films. *Applied Physics Letters*, 66(10):1166, 1995. ISSN 00036951.

[75] M. Müllenborn, H. Dirac, and J. W. Petersen. Silicon nanostructures produced by laser direct etching. *Applied Physics Letters*, 66(22):3001, 1995. ISSN 00036951.

[76] Y. Xia, J. J. McClelland, R. Gupta, D. Qin, X.-M. Zhao, L. L. Sohn, R. J. Celotta, and G. M. Whitesides. Replica molding using polymeric materials: A practical step toward nanomanufacturing. *Advanced Materials*, 9(2):147–149, February 1997. ISSN 1521-4095.

[77] U. Plachetka, M. Bender, A. Fuchs, B. Vratzov, T. Glinsner, F. Lindner, and H. Kurz. Wafer scale patterning by soft UV-Nanoimprint lithography. *Microelectronic Engineering*, 73-74:167–171, June 2004. ISSN 0167-9317.

[78] Y. Xia, J. Tien, D. Qin, and G. M. Whitesides. Non-Photolithographic methods for fabrication of elastomeric stamps for use in microcontact printing. *Langmuir*, 12(16):4033–4038, January 1996.

[79] X.-M. Zhao, Y. Xia, and G. M. Whitesides. Fabrication of three-dimensional micro-structures: Microtransfer molding. *Advanced Materials*, 8(10):837–840, October 1996. ISSN 1521-4095.

[80] G. M. Whitesides, E. Ostuni, S. Takayama, X. Jiang, and D. E. Ingber. Unconventional nanofabrication. *Annual Review of Biomedical Engineering*, 3(1):335–373, 2001. ISSN 1523-9829.

REFERENCES

[81] G. S Fiorini and D. T Chiu. Disposable microfluidic devices: fabrication, function, and application. *BioTechniques*, 38(3):429–446, March 2005. ISSN 0736-6205. PMID: 15786809.

[82] J. R. Anderson, D. T. Chiu, R. J. Jackman, O. Cherniavskaya, J. C. McDonald, H. Wu, S. H. Whitesides, and G. M. Whitesides. Fabrication of topologically complex Three-Dimensional microfluidic systems in PDMS by rapid prototyping. *Analytical Chemistry*, 72(14):3158–3164, July 2000.

[83] Y. Zhang, C.-W. Lo, J. A. Taylor, and S. Yang. Replica molding of High-Aspect-Ratio polymeric nanopillar arrays with high fidelity. *Langmuir*, 22(20):8595–8601, 2006.

[84] J. Feng, B. Huang, and M. Zhong. Fabrication of superhydrophobic and heat-insulating antimony doped tin oxide/polyurethane films by cast replica micromolding. *Journal of Colloid and Interface Science*, 336(1):268–272, August 2009. ISSN 0021-9797.

[85] D. Cai and A. Neyer. Cost-effective and reliable sealing method for pdms (polydimethylsiloxane)-based microfluidic devices with various substrates. *Microfluidics and Nanofluidics*, Volume 9, Numbers 4-5:855–864, 2010.

[86] Datasheet Elastosil RT 601, 2010.

[87] H. Schmitt, M. Rommel, A.J. Bauer, L. Frey, A. Bich, M. Eisner, R. Voelkel, and M. Hornung. Full wafer microlens replication by UV imprint lithography. *Microelectronic Engineering*, 87(5-8):1074–1076, May 2010. ISSN 0167-9317.

[88] R. Ji, M. Hornung, M. A. Verschuuren, R. van de Laar, J. van Eekelen, U. Plachetka, M. Moeller, and C. Moormann. UV enhanced substrate conformal imprint lithography (UV-SCIL) technique for photonic crystals patterning in LED manufacturing. *Microelectronic Engineering*, 87(5-8):963–967, May 2010. ISSN 0167-9317.

[89] F. van Delft, R. van de Laar, M. Verschuuren, E. Platzgummer, and H. Loeschner. Charged particle nanopatterning (CHARPAN) of 2D and 3D masters for flexible replication in substrate conformal imprint lithography (SCIL). *Microelectronic Engineering*, 87(5-8):1062–1065, May 2010. ISSN 0167-9317.

[90] H. Schmid and B. Michel. Siloxane polymers for High-Resolution, High-Accuracy soft lithography. *Macromolecules*, 33(8):3042–3049, 2000. ISSN 0024-9297.

[91] S. Bhattacharya, Y. Gao, V. Korampally, M.T. Othman, S. A. Grant, K. Gangopadhyay, and S. Gangopadhyay. Mechanics of plasma exposed spin-on-glass (SOG) and polydimethyl siloxane (PDMS) surfaces and their impact on bond

strength. *Applied Surface Science*, 253(9):4220–4225, February 2007. ISSN 0169-4332.

[92] T. W. Odom, J. C. Love, D. B. Wolfe, K. E. Paul, and G. M. Whitesides. Improved pattern transfer in soft lithography using composite stamps. *Langmuir*, 18(13):5314–5320, June 2002.

[93] H. Kang, J. Lee, J. Park, and H. H. Lee. An improved method of preparing composite poly(dimethylsiloxane) moulds. *Nanotechnology*, 17(1):197–200, 2006. ISSN 0957-4484.

[94] D. Losic, J.G. Mitchell, R. Lal, and N.H. Voelcker. Rapid fabrication of micro- and nanoscale patterns by replica molding from diatsom biosilica. *Advanced Functional Materials*, 17:2439–2446, 2007.

[95] R.A. Singh, E.-S. Yoon, H.J. Kim, J. Kim, H.E. Jeong, and K.Y. Suh. Replication of surfaces of natural leaves for enhanced micro scale tribologial property. *Material Science and Engineering C*, 27:875–879, 2007.

[96] M. Sun, C. Luo, L. Xu, H. Ji, Q. Ouyang, D. Yu, and Y. Chen. Artificial lotus leaf by nanocasting. *Langmuir*, 21:8978–8981, 2005.

[97] M. Beck, M. Graczyk, I. Maximov, E. -L. Sarwe, T. G. I. Ling, M. Keil, and L. Montelius. Improving stamps for 10 nm level wafer scale nanoimprint lithography. *Microelectronic Engineering*, 61-62:441–448, July 2002. ISSN 0167-9317.

[98] H. Schift, S. Saxer, S. Park, C. Padeste, U. Pieles, and J. Gobrecht. Controlled co-evaporation of silanes for nanoimprint stamps. *Nanotechnology*, 16(5):S171–S175, 2005. ISSN 0957-4484.

[99] *Bedienungsanleitung Heißprägeanlage HEX03, Jenoptik Laser, Optik, Systeme GmbH, 2001*.

[100] I. Fernández Cuesta. *Nanoimprint Lithography: Developments and nanodevice fabrication*. PhD thesis, Universitat Autonoma de Barcelona, 2009.

[101] T Senn, J P Esquivel, M Lörgen, N Sabaté, and B Löchel. Replica molding for multilevel micro-/nanostructure replication. *Journal of Micromechanics and Microengineering*, 20(11):115012, 2010. ISSN 0960-1317.

[102] J.W.S. Rayleigh. On the maintenance of vibrations by forces of double frequency, and on the propagation of waves through a medium endowed with periodic structure. *Philosophical Magazine*, 24:145–159, 1887.

REFERENCES

[103] K. J. Knopp, D. H. Christensen, J. R. Hill, and K. D. Masterson. Thin-film design for enhanced stability of optically pumped vertical-cavity surface-emitting lasers (VCSELs). *Surface and Coatings Technology*, 86-87(Part 2):783–787, December 1996. ISSN 0257-8972.

[104] M. S. Alias, S. Shaari, P. O. Leisher, and K. D. Choquette. Single transverse mode control of VCSEL by photonic crystal and trench patterning. *Photonics and Nanostructures - Fundamentals and Applications*, 8:38–46, 2010. ISSN 1569-4410.

[105] S. John. Strong localization of photons in certain disordered dielectric superlattices. *Physical Review Letters*, 58(23):2486, June 1987.

[106] W. Bogaerts, D. Taillaert, B. Luyssaert, P. Dumon, J. Van Campenhout, P. Bienstman, D. Van Thourhout, R. Baets, V. Wiaux, and S. Beckx. Basic structures for photonic integrated circuits in silicon-on-insulator. *Optics Express*, 12(8):1583–1591, April 2004.

[107] S. McNab, N. Moll, and Y. Vlasov. Ultra-low loss photonic integrated circuit with membrane-type photonic crystal waveguides. *Optics Express*, 11(22):2927–2939, November 2003.

[108] Y. A. Vlasov, M. O'Boyle, H. F. Hamann, and S. J. McNab. Active control of slow light on a chip with photonic crystal waveguides. *Nature*, 438(7064):65–69, November 2005. ISSN 0028-0836.

[109] D. Erickson, T. Rockwood, T. Emery, A. Scherer, and D. Psaltis. Nanofluidic tuning of photonic crystal circuits. *Optics Letters*, 31(1):59–61, January 2006.

[110] A. Gomyo, J. Ushida, and M. Shirane. Highly drop-efficient channel-drop optical filters with si-based photonic crystal slabs. *Thin Solid Films*, 508(1-2):422–425, June 2006. ISSN 0040-6090.

[111] H.-T. Chien, C.-C. Chen, and P.-G. Luan. Photonic crystal beam splitters. *Optics Communications*, 259(2):873–875, March 2006. ISSN 0030-4018.

[112] P.F. Xing, P.I. Borel, L.H. Frandsen, A. Harpoth, and M. Kristensen. Optimization of bandwidth in 60° photonic crystal waveguide bends. *Optics Communications*, 248(1-3):179–184, April 2005. ISSN 0030-4018.

[113] M. Loncar, T. Yoshie, A. Scherer, P. Gogna, and Y. Qiu. Low-threshold photonic crystal laser. *Applied Physics Letters*, 81(15):2680–2682, 2002.

[114] E. Waks, D. Englund, D. Fattal, J. Vuckovic, and Y. Yamamoto. Photonic-crystal based single photon source. volume 5893, pages 173–186, August 2005.

REFERENCES

[115] J. Kouba, S. Kiss, M. Barth, W. Eberhardt, and B. Loechel. Fabrication and optical characterization of si_3n_4 2D-photonic crystals for applications in visible range. In Elizabeth A. Dobisz and Louay A. Eldada, editors, *Nanoengineering: Fabrication, Properties, Optics, and Devices IV*, volume 6645, pages 664505–9, San Diego, CA, USA, 2007. SPIE.

[116] Ch.-G. Choi, Y.-T. Han, J.T. Kim, and H. Schift. Air-suspended two-dimensional polymer photonic crystal slab waveguides fabricated by nanoimprint lithography. *Applied Physics Letters*, 90(22):221109–221109–3, May 2007. ISSN 10773118.

[117] P. I. Borel, B. Bilenberg, L. H. Frandsen, T. Nielsen, J. Fage-Pedersen, A. V. Lavrinenko, J. S. Jensen, O. Sigmund, and A. Kristensen. Imprinted silicon-based nanophotonics. *Optics Express*, 15:1261–1266, 2007.

[118] J. Kouba. *Investigation of silicon nitride based two-dimensional photonic crystals for the visible spectral range*. PhD thesis, Technische Universität Berlin, 2008.

[119] N. Nüsse. *Hybridstrukturen aus Nanodiamanten, nanoplasmonischen Elementen und photonischen Kristallen*. PhD thesis, Technische Universität Berlin, 2011.

[120] M. Barth. *Hybrid Nanophotonic Elements and Sensing Devices based on Photonic Crystal Structures*. PhD thesis, Humboldt-Universität zu Berlin, 2010.

[121] R. Truckenmüller. *Herstellung von dreidimensionalen Mikrostrukturen aus Polymermembranen*. PhD thesis, Universtität Karlsruhe, 2003.

[122] H. Becker and U. Heim. Hot embossing as a method for the fabrication of polymer high aspect ratio structures. *Sensors and Actuators A: Physical*, 83(1-3):130–135, May 2000. ISSN 0924-4247.

[123] Y. Hirai, S. Yoshida, S. Takagi, Y. Tanaka, H. Yabe, K. Sasaki, and H. Sumitani. High aspect pattern fabrication by nano imprint lithography using fine diamond mold. *Microprocesses and Nanotechnology Conference, 2002. Digest of Papers. Microprocesses and Nanotechnology 2002. 2002 International*, pages 26 – 27, 2002.

[124] M.R. Cardoso, V. Tribuzi, D.T. Balogh, L. Misoguti, and C.R. Mendonça. Laser microstructuring for fabricating superhydrophobic polymeric surfaces. *Applied Surface Science*, 257(8):3281–3284, February 2011. ISSN 0169-4332.

[125] N. J. Shirtcliffe, S. Aqil, C. Evans, G. McHale, M. I. Newton, C. C. Perry, and P. Roach. The use of high aspect ratio photoresist (SU-8) for super-hydrophobic pattern prototyping. *Journal of Micromechanics and Microengineering*, 14(10): 1384–1389, 2004. ISSN 0960-1317.

REFERENCES

[126] C. I. Park, H. E. Jeong, S. H. Lee, H. S. Cho, and K. Y. Suh. Wetting transition and optimal design for microstructured surfaces with hydrophobic and hydrophilic materials. *Journal of Colloid and Interface Science*, 336(1):298–303, August 2009. ISSN 0021-9797.

[127] H.E. Jeong, M.K. Kwak, C.I. Park, and K.Y. Suh. Wettability of nanoengineered dual-roughness surfaces fabricated by uv-assisted force lithography. *Journal of Colloid and Interface Science*, 339:202–207, 2009.

[128] X. Liu and C. Luo. Fabrication of super-hydrophobic channels. *Journal of Micromechanics and Microengineering*, 20(2):025029, 2010. ISSN 0960-1317.

[129] Peter Steinke. *Finite-Elemente-Methode*. Springer-Verlag, 2007.

[130] T.Senn, C. Müller, and H. Reinecke. Replication of HARMST and large area nanostructured parts using UV cationic polymerization. *Journal of Micromechanical Microengineering*, 20:075002, 2010.

[131] J.-C. Roulet, R. Völkel, H. P. Herzig, E. Verpoorte, N. F. de Rooij, and R. Dändliker. Performance of an integrated microoptical system for fluorescence detection in microfluidic systems. *Analytical Chemistry*, 74(14):3400–3407, 2002.

[132] Y. Sun and S. R. Forrest. Organic light emitting devices with enhanced outcoupling via microlenses fabricated by imprint lithography. *Journal of Applied Physics*, 100(7):073106, 2006. ISSN 00218979.

[133] R. Völkel. Halbleitertechnik für bessere Mobiltelefon-Kameras. *Mikroproduktion*, 01/09:52–56, 2009.

[134] M. Hennemeyer, M. Hornung, R. Zoberbier, M. Eisner, R. Völkel, and D. Tönnies. Soft Imprint Lithographie für die Herstellung von Mikrolinsen. In *5. Dornbirner Mikrotechniktage*, 2010.

[135] H. Han and K. Main. Low cost camera modules using integration of waferscale optics and wafer-level packaging of image sensors. pages 76311P–76311P–7, Shanghai, China, 2009.

[136] S. S. Oh, C.-G. Choi, and Y.-S. Kim. Fabrication of micro-lens arrays with moth-eye antireflective nanostructures using thermal imprinting process. *Microelectronic Engineering*, 87(11):2328–2331, 2010. ISSN 01679317.

[137] S. Ziokowski, I. Frese, H. Kasprzak, and S. Kufner. Contactless embossing of microlenses - a parameter study. *Optical Engineering*, 42(5):1451, 2003. ISSN 00913286.

REFERENCES

[138] M. Rossi, H. Rudmann, S. Westenhöfer, M. Salt, and R. Pelzer. Optical module fabrication using nanoimprint technology. In *Micromachining Technology for Micro-Optics and Nano-Optics IV*, volume 6110, pages 61100L–61100L–3. SPIE, 2006.

[139] M.-C. Chou, C.T. Pan, S.C. Shen, M.-F. Chen, K.L. Lin, and S.-T. Wu. A novel method to fabricate gapless hexagonal micro-lens array. *Sensors and Actuators A: Physical*, 118(2):298–306, February 2005. ISSN 0924-4247.

[140] C.T. Pan and C.H. Su. Fabrication of gapless triangular micro-lens array. *Sensors and Actuators A: Physical*, 134(2):631–640, March 2007. ISSN 0924-4247.

[141] K.H. Liu, M.F. Chen, C.T. Pan, M.Y. Chang, and W.Y. Huang. Fabrication of various dimensions of high fill-factor micro-lens arrays for OLED package. *Sensors and Actuators A: Physical*, 159(1):126–134, April 2010. ISSN 0924-4247.

[142] H. Rudmann and M. Rossi. Design and fabrication technologies for ultraviolet replicated micro-optics. *Optical Engineering*, 43:2575, 2004. ISSN 00913286.

[143] U. Schulz. Review of modern techniques to generate antireflective properties on thermoplastic polymers. *Applied Optics*, 45(7):1608–1618, March 2006. ISSN 0003-6935. PMID: 16539270.

[144] S. Chattopadhyay, Y.F. Huang, Y.J. Jen, A. Ganguly, K.H. Chen, and L.C. Chen. Anti-reflecting and photonic nanostructures. *Materials Science and Engineering: R: Reports*, 69(1-3):1–35, June 2010. ISSN 0927-796X.

[145] A. Gombert, W. Glaubitt, K. Rose, J. Dreibholz, C. Zanke, B. Bläsi, A. Heinzel, W. Horbelt, D. Sporn, W. Döll, V. Wittwer, and J. Luther. Glazing with very high solar transmittance. *Solar Energy*, 62:177–188, 1998.

[146] A. Gombert, W. Glaubitt, K. Rose, J. Dreibholz, B. Bläsi, A. Heinzel, D. Sporn, W. Döll, and V. Wittwer. Antireflective transparent covers for solar devices. *Solar Energy*, 68:357–360, 2000.

[147] C. Bühler. *Mikrostrukturen zur Streuung von Tageslichtstroemen*. PhD thesis, Albert-Ludwigs-Universität Freiburg i. Br., 2003.

[148] B. Bläsi. *Holographisch hergestellte Antireflexoberflächen für solare und visuelle Anwendungen*. PhD thesis, Albert-Ludwigs-Universität Freiburg i. Br., 2000.

[149] J.P. Esquivel, T. Senn, P. Hernandez-Fernandez, J. Santander, M. Lörgen, S. Rojas, B. Löchel, C. Cané, and N. Sabaté. Towards a compact SU-8 micro-direct methanol fuel cell. *Journal of Power Sources*, 195(24):8110–8115, December 2010. ISSN 0378-7753.

[150] N.-T. Nguyen and S. H. Chan. Micromachined polymer electrolyte membrane and direct methanol fuel cells a review. *Journal of Micromechanics and Microengineering*, 16(4):R1–R12, 2006. ISSN 0960-1317.

[151] A. Kundu, J.H. Jang, J.H. Gil, C.R. Jung, H.R. Lee, S.-H. Kim, B. Ku, and Y.S. Oh. Micro-fuel cells–Current development and applications. *Journal of Power Sources*, 170(1):67–78, June 2007. ISSN 0378-7753.

[152] S.K. Kamarudin, W.R.W. Daud, S.L. Ho, and U.A. Hasran. Overview on the challenges and developments of micro-direct methanol fuel cells (DMFC). *Journal of Power Sources*, 163(2):743–754, January 2007. ISSN 0378-7753.

[153] J. P. Esquivel Bojorquez. *Microfabricated Fuel Cell as Power Sources for MEMS*. PhD thesis, Universitat Autonoma de Barcelona, 2011.

[154] R. Chen and T.S. Zhao. Performance characterization of passive direct methanol fuel cells. *Journal of Power Sources*, 167(2):455–460, May 2007. ISSN 0378-7753.

[155] F. A. de Bruijn, D. C. Papageorgopoulos, E. F. Sitters, and G. J. M. Janssen. The influence of carbon dioxide on PEM fuel cell anodes. *Journal of Power Sources*, 110(1):117–124, July 2002. ISSN 0378-7753.

[156] W.-M. Yan, H.-S. Chu, M.-X. Lu, F.-B. Weng, G.-B. Jung, and C.-Y. Lee. Degradation of proton exchange membrane fuel cells due to CO and CO2 poisoning. *Journal of Power Sources*, 188(1):141–147, March 2009. ISSN 0378-7753.

[157] O.Mertsch, D.Schondelmaier, I.Rudolph, O.Kutz, A. D. Walter, A.Schleunitz, J. Kouba, Ch. Waberski, and B.Löchel. Generation and characterization of superhydrophobic micro- and nano-structured surfaces. *Journal of Adhesion Science and Technology*, 22:1967–1983, 2008.

[158] O. Mertsch. *Nanoporöse und ultra-hydrophobe Strukturen im Negativresist SU8*. PhD thesis, Technische Universität Berlin, 2008.

[159] F.J. Blanco, M. Agirregabiria, J. Garcia, J. Berganzo, M. Tijero, M.T. Arroyo, J.M. Ruano, I. Aramburu, and K. Mayora K. Novel three-dimensional embedded su-8 microchannels fabricated using a low temperature full wafer adhesive bonding. *Journal of Micromechanics and Microengineering*, 14:1047–1056, 2004.

Appendix

Stamps for HE-NIL (Process 1)

	Process Step	Technical Parameters	Remarks
1.1	**Resist Coating**		
1.1.1	**Pretreatment of the Substrate**	Standard Si substrate, Si substrate, 4", <100>, thickness d = 525 μm, one side polished, dehydration in a vacuum oven at 200 °C	
1.1.2	**Coating Resist and Post Bake**	No priming necessary, PMMA (2.2M), Micro Chem Corp., USA speed: 1500 rpm time: 45 s thickness: 140 nm bake 20 min @ 180°C (convection oven)	Alternative: bake 1 min @ 170 °C (hot plate)
1.2	**Electron Beam Lithography**		
1.2.1	**Design Preparation**	preparation of the pattern design using a CAD program for further processing in the software of the EBL writing system. Pattern optimization in the EBL software for writing time reduction, proximity corrections (long and short range corrections)	stamps for nanoimprint can be optimized by minimizing the elevated structures (reduction of imprint time). In order to obtain also short writing times, the optimal approach is to use either a negative resist for direct pattern transfer or a positive resist and transfer the structures using a hard mask, structured by lift-off.
1.2.2	**Pattern Definition**	Serial exposure by focused ion beam, PMMA exposure using a 100 kV electron beam. Dose: 800 μC/cm^2	

1.2.3	Resist Development	Wet development in AR 600-50 for 10 s AR 600-60 for 10 s rinsed in IPA for 10 s	
1.3	**Pattern Transfer**		
1.3.1	Deposition of Hard Mask	Evaporation of a 10 - 20 nm thick Ni layer	Alternative: Cr can be used as hard mask
1.3.2	Lift Off	Lift-off in N,N-dimethylformamid (DMF) to create a hard mask. Leave substrate in DMF for several hours, finishing using an ultrasonic step, rinse in water	In contrast to the dissolution of PMMA in acetone, DMF results in a moisture expansion which favours the lift-off, as the material on the sidewall of the structure is lifted as well, resulting in a clean lift-off.
1.3.3	Reactive Ion Etching	Dry etching of silicon to obtain vertical sidewalls. The etch process uses a combination of gases (e.g. SF_6 as etch gas and C_4F_8 as passivation gas)	Undercuts and roughness have to be avoided, because this results in enhanced demolding forces and damage of structures in NIL.
1.3.4	Mask Removing	The Ni mask can be removed using a HCl dip for 20 min	If Cr is used as mask material a combination of $HClO_4$ and $(NH_4)_2$ [Ce $(NO_3)_6$] can be used to remove the mask after etching.

Appendix

1.4	**Anti-Adhesive Coating**		
1.4.1	**Preparation of the Stamp Surface**	Cleaning and activation by RIE process using O_2 as etch gas to remove all organic residue and to activate the surface (generation of free reactive silanol bonds for silane binding). RF power 50 W, O_2 flow of 60 sccm for 60 s	Alternatively the cleaning and activation of the surface can be done using a solution of H_2O_2 and H_2SO_4
1.4.2	**Chemical Vapour Deposition**	Deposition of anti-sticking layer. A monolayer of F13-TCS is deposited from the vapour phase in a convection oven @ 200 °C for several hours	Alternatively an anti sticking layer can be deposited by dip coating in a 1-10 mM solution of perfluorotrichlorsilane for 1-2 hours.

Stamps for UV-NIL (Process 2)

	Process Step	Technical Parameters	Remarks
2.1	**Master Structure**		
2.1.1	**Master Fabrication**	Depending on the structure design, different processes for the master fabrication can be used including e.g. UV Lithography, Interference Lithography or EBL.	The resolution and the aspect ratio that can be replicated depends on the mechanical properties of the PDMS material. This has to be taken into account in the structure design of the master.
2.2	**PDMS Preparation**		
2.2.1	**Preparation the Prepolymer**	The 2 component PDMS materials (Elastosil 601 A and B) are mixed in a 9:1 proportion (base material: catalyst) and degassed for about 10 min.	The standard PDMS material is Sylgard 184 PDMS (Dow Corning).
2.3	**PDMS Stamp Fabrication**		
2.3.1	**Casting of PDMS**	The liquid PDMS prepolymer mixture is cast onto the master covering the surface structure.	To obtain a thin PDMS stamp the polymer can also be spincoated at low spin velocities.
2.3.2	**Degassing and Curing**	The covered master is degassed for a few minutes to obtain a complete form filling of the structures. The PDMS is cured for 30 min @ 70 °C.	
2.3.3	**Peel off the Stamp**	For Elastosil 601 the stamp can easily be peeled off the master structure.	For other PDMS materials it can be necessary to deposit an anti-sticking layer on the master to peel off the structure without damages. This also depends on the material of the master (e.g. resist, silicon, metal)

Appendix

HE-NIL process (Process 3)

	Process Step	Technical Parameters	Remarks
3.1	Resist Preparation and Coating		
3.1.1	Preparation of the Resist Solution	Dissolving PMMA in chlorobenzene to the desired viscosity.	Alternative: commercially available resist materials can be used.
3.1.2	Resist Coating and Post Bake	Spincoating of the PMMA solution. The speed and time depends on the viscosity of the PMMA solution. The thickness can range between some tens of nanometres to various micrometres. bake 20 min @ 180 °C	For commercially available materials spin curves and baking times can be obtained from the data sheets.
3.2	Imprint Process		
3.2.1	Substrate Stamp Align	The stack consisting of the substrate, the stamp and a compliance layer is pre-assembled in the hot embossing machine prior to heating and pressing. A small force is used to fix the assembly and to obtain a better thermal contact.	If sticking problem between silicon and compliance layer and between silicon or compliance layer to the embossing plates occur, a PI (polyimid) foil can be introduced to reduce the adhesion.
3.2.2	Pattern Transfer into Resist	The imprint into the PMMA layer is typically carried out at a temperature of 190 °C and a pressure of 10 - 15 MPa. The embossing time depends on the stamp design.	
3.2.3	Demolding	The pressure is released at a temperature of 60 °C. The demolding is done manually by introducing a razor plate between stamp and substrate.	

145

3.2.4	Etching of Residual Layer	Etching of residual layer using RIE with an oxygen plasma. oxygen flow: 20 sccm RF power: 20 W pressure: 75 mTorr etch rate: 32 nm/min	
3.3	**Direct Pattern Transfer**		
3.3.1	RIE Etching with PMMA Mask	RIE process using PMMA as etch mask with CHF_3 selectivity of Si to PMMA : 2	For the direct pattern transfer using the structured PMMA layer as an etch mask, the parameters have to be optimized in order to obtain a high selectivity. A higher selectivity can be obtained by cooling the substrate (cryo etching)
3.3.2	Removal of PMMA	RIE resist ashing with an oxygen plasma for a few seconds allows to remove the resist without affecting the silicon surface.	The resist can also be removed using wet etching with e.g. acetone
3.4	**Pattern Transfer via Lift Off**		
3.4.1	Deposition of Hard Mask	A 10 - 20 nm thick Ni mask is deposited using evaporation	Alternative Cr can be used as etch mask
3.4.2	Lift Off	Lift-off in N,N-dimethylformamid (DMF) to create a hard mask. The substrate is left in DMF for several hours. Finishing using an ultrasonic step. Rinse in water	In contrast to the dissolution of PMMA in acetone, DMF results in a moisture expansion which favours the lift-off, as the material on the sidewall of the structure is lifted as well, resulting in a clean lift-off

3.4.3	**Reactive Ion Etching**	Dry etching of silicon to obtain vertical sidewalls. The etch process uses a combination of gases (e.g. SF_6 as etch gas and C_4F_8 as passivation gas)	In comparison to the direct pattern transfer, the use of a hard mask offers a higher selectivity in the etching process, thus higher aspect ratios can be obtained. Furthermore the tone of the replication can be changed
3.4.4	**Mask Removing**	The Ni mask can be removed using a HCl dip for 20 min	If Cr is used as mask material a combination of $HClO_4$ and $(NH_4)_2\,[Ce\,(NO_3)_6]$ can be used to remove the mask after etching.

UV-NIL Process (Process 4)

	Process Step	Technical Parameters	Remarks
4.1	**Resist Preparation and Coating**		
4.1.1	**Preparation of Resist Solution**	An epoxy mixture is used as resist material in the UV-NIL process. Different monomers can be used (e.g. BCE, DGE-BPA, DGE-TEG). The monomer solution is mixed with 3 wt-% of the photoinitiator (TAS)	The composition of the resist material has influence on the chemical and physical behaviour (e.g. chemical resistance, viscosity etc.)
4.1.2	**Coating Resist**	Spincoating of resist material. Depending on the monomer mixture (viscosity) and the desired final thickness of the resist material, the speed during spincoating has to be adjusted.	The viscosity ranges between 20 and 130 mPas for the monomer materials used in this work.
4.2	**Imprint Process**		
4.2.1	**Substrate Stamp Align**	The stamp and the coated substrate are aligned to each other.	The stamp should not touch the coated substrate in the alignment process, since a low viscosity material is used and cavity filling would occur.
4.2.2	**Pattern Transfer into Resist**	Stamp and coated substrate are brought into contact. The epoxy material is cured using UV exposure.	The resist materials in UV-NIL have much lower viscosities compared to those used in HE-NIL. Therefore, cavity filling can occur without pressure. The stamp cavities are filled within seconds. Stamp and substrate have to be brought in contact carefully to avoid air bubbles in between.

Appendix

4.2.3	Demolding	The demolding process can be done easily by peeling the PDMS stamp off the cured epoxy layer.	The UV-NIL process was used to replicate permanent structures onto a substrate. Therefore the residual layer is not removed.

Further processing of the structured master as e.g. lift-off is difficult due to the crosslinking after curing

Thermoforming Process (Process 5)

	Process Step	Technical Parameters	Remarks
5.1	**Foil Preparation**		
5.1.1	**Deposition of PDMS Membrane**	A thin membrane of PDMS is spincoated onto the backside of the foil with a speed of 5000 rpm resulting in a elastic membrane of 10 μm thickness. The PDMS material is cured @ 70 °C for 30 min.	The PDMS membrane stabilizes the foil in the following Thermoforming process allowing for the use of higher pressure and temperature in the process.
5.1.1	**Preparation of Resist Solution**	The preparation of the resist solution is done analogue to the UV-NIL process (see 4.1.1)	In this application the solution must contain a minimal amount of ether (DGE-TEG) to obtain sufficient sticking of the layer to the foil after structuring.
5.1.2	**Resist Coating**	The resist material is spincoated on top of the foil the speed depends on the desired final thickness of the layer	The thickness of the layer has to be adjusted to the depth of the structures to obtain a minimal residual layer thickness. A too thick layer can crack during the Thermoforming process due to the applied pressure and temperature. The viscosity of the monomer mixture can be adjusted by the amount of ether.
5.2	**Foil Structuring**		
5.2.1	**Master Fabrication**	The technology for the master fabrication depends on the structures that are desired to integrate.	Depending on the size and shape processes like UV Lithography, EBL or Interference Lithography can be used.

5.2.2	PDMS Stamp Fabrication	The PDMS stamp fabrication is analogue to the process steps described in 2.3.	
5.2.3	Peel Off the PDMS Stamp	For Elastosil 601 the stamp can easily be peeled off the master structure	For other PDMS materials it can be necessary to deposit an anti sticking layer on the master to peel off the structure without damages. This also depends on the material of the master (e.g. resist, silicon, metal)
5.2.4	Pattern Transfer	Stamp and coated substrate are brought into contact. The epoxy material is cured using UV exposure.	The resist materials in UV-NIL have much lower viscosity compared to those used in HE-NIL. Therefore, cavity filling can occur without pressure. The stamp cavities are filled within seconds. Stamp and substrate have to be brought in contact carefully to avoid air bubbles in between.
5.2.5	Demolding	The demolding process can be done easily by peeling the PDMS stamp off the cured epoxy layer	

5.3	Thermoforming		
5.3.1	Fabrication of the Mold Insert	The fabrication of the mold insert can be performed using different process like lithography processes or micro mechanical processes e.g. micro milling	The mold insert contains the microstructures where the surface structures of the foil should be applied to.
5.3.2	Substrate Stamp Align in Thermoforming Tool	The structured foil is aligned to the microstructures in the mold insert. The aligned structured foil is clamped using the ring in the Thermoforming tool and the upper part is screwed onto the ring to close the tool.	The integrated ring is also used for the fabrication of 3D structured PDMS stamps (see also point 6)
5.3.3	Structure Integration	The entire Thermoforming tool is heated up above the glass transition temperature of the foil and then pressure is applied through the upper part leading to the foil covering the microstructured mold insert. The tool is then cooled down again and the pressure is released from the tool.	The deposition of the PDMS membrane on the backside of the foil allows the use of higher temperature and pressure in the Thermoforming process leading to a higher molding accuracy
5.3.4	Detach Foil from Tool	The structured foil can be detached from the mold insert by removing the upper part and the ring of the tool.	If a 3D structured PDMS stamp is produced from the foil, the foil maintains clamped between the lower part and the ring and the PDMS prepolymer is cast on top of the structured foil

Appendix

Replica Molding (Process 6)

	Process Step	Technical Parameters	Remarks
6.1	**PDMS Stamp Fabrication**		
6.1.1	**Master fabrication**	The master fabrication for the PDMS stamps can be done using different technologies e.g. a 3D structured foil from the Thermoforming can be used	
6.1.2	**Casting of PDMS**	The PDMS stamp fabrication is analogue to the process steps described in 2.3	
6.1.3	**Peel Off the Stamp**	For Elastosil 601 the stamp can easily be peeled off the master structure	For other PDMS materials it can be necessary to deposit an anti-sticking layer on the master to peel off the structure without damages. This also depends on the material of the master (e.g. resist, silicon, metal)
6.2	**Replications**		
6.2.1	**Resist Preparation**	The resist materials that were used in the Replica Molding process are the same as mentioned in 4.1.1	In this application rather thick polymer parts are produced than thin films. Therefore the ether component is not important for these application
6.2.2	**Casting of Resist Material**	The monomer mixture is cast onto the PDMS stamp and excess material is removed by applying a polymer foil on top.	The monomer material fills the cavities in the stamp without special treatment.

153

6.2.3	**Curing of Resist**	The resist material is cured by UV exposure. The dose depends on the thickness of the part. For a part with a thickness of 1 mm a dose of 1200 $\frac{mJ}{cm^2}$ is enough to demold it without damages.	In cationic polymerization mechanism as it applies for the materials used here, after the exposure the polymerization reaction continues and therefore the exposure time can be kept short.
6.2.4	**Demolding**	The cured polymer part is peeled off the PDMS stamp.	

i want morebooks!

Buy your books fast and straightforward online - at one of the world's fastest growing online book stores! Environmentally sound due to Print-on-Demand technologies.

Buy your books online at
www.get-morebooks.com

Kaufen Sie Ihre Bücher schnell und unkompliziert online – auf einer der am schnellsten wachsenden Buchhandelsplattformen weltweit!
Dank Print-On-Demand umwelt- und ressourcenschonend produziert.

Bücher schneller online kaufen
www.morebooks.de

OmniScriptum Marketing DEU GmbH
Heinrich-Böcking-Str. 6-8
D - 66121 Saarbrücken
Telefax: +49 681 93 81 567-9

info@omniscriptum.de
www.omniscriptum.de

Printed by Books on Demand GmbH, Norderstedt / Germany